DO YOUR OWN

PROFESSIONAL WELDING

Other TAB books by the author:

No. 1384
$16.95

DO YOUR OWN

PROFESSIONAL WELDING

BY CHARLES R. SELF

 TAB BOOKS Inc.
BLUE RIDGE SUMMIT, PA. 17214

For Caroline, who has seen this book and 17 others come and go with her help, and who will see all the rest.

FIRST EDITION

FIRST PRINTING

Library of Congress Cataloging in Publication Data

Self, Charles R.
    Do your own professional welding.

    Includes index.
    1. Welding—Amateurs' manuals.   I. Title.
TS227.S416   671.5′2   81-18371
ISBN 0-8306-0068-X   AACR2
ISBN 0-8306-1384-6 (pbk.)

Cover photo courtesy of Airco Welding Products, Murray, New Jersey.

Contents

Introduction

For many years welding and brazing have been considered provinces of the true professional or the most finicky hobbyist. This is no longer true. As more light welding equipment is developed, the work is more within the reach of a person with decent eye and hand coordination.

Sears now has a lightweight 100-ampere arc welder, and Montgomery Ward has a 50-ampere unit. Bernzomatic produces a lightweight oxypropane torch, and Solidox has another oxypropane torch that uses an oxygen "candle" instead of bottled oxygen. A lightweight welding outfit such as Airco's Tote-Weld oxy-MAPP gas unit is more expensive and much more versatile. This unit provides professional style and accuracy with lighter capacity and is easily movable. Oxyacetylene torches are available for welding, brazing, and cutting. Arc welders have capacities of at least 300 amperes.

The home welder has more tools to choose from than ever before. Base prices for the smallest units start at about $30, though top-of-the-line models can cost more than $500. You can select an ac-dc welder (use of reverse polarity dc makes certain welding jobs easier), add a tungsten inert gas (TIG) accessory kit, and weld similar metals. A high frequency stabilizer with the TIG accessories gives the same capacity to an ac arc welder. Bothersome work for arc welders, such as cutting and brazing, can be handled with carbon arc torches.

Companies such as Lincoln Electric make phenomenal home-size welders. Smaller companies like Century provide a smaller variety, but they still may offer as many as 15 different machines. For the novice, this choice is bound to be confusing, especially when you have to include the gas welding and brazing equipment made by Airco, Harris Calorific, and Dockson.

Welding is not a skill you can learn overnight. It simply is not possible to pick up a torch or electrode holder and run a good bead without practice. A solid understanding of the basics of what to do, along with correct welding rod (or electrode) and welder selection, should allow anyone with normal coordination to learn to weld reasonably well with two or three weeks of daily practice. You still won't be a professional, and there will be many jobs you can't even consider doing. You will be able to run basic beads and weaves.

Brazing, or more correctly braze welding, is even simpler than fusion welding, at least regarding rod selection and application of heat. A bit less time is needed for practice. Brazing and braze welding are among the handiest skills you can have. I recently finished using Airco's Tote-Weld to repair cables for an overhead garage door, as the particular clamp needed is no longer made. The job couldn't have been done without braze welding, short of replacing the entire cable system. Brazing is useful if you are assembling some types of solar systems, too. Used for joints, brazing provides a leakfree and long lasting method of joining copper tubing or pipe. For engine block repairs, braze welding usually makes a stronger union than fusion welding because of the cast iron's high carbon content.

Soldering can be the easiest way to join metal by using heat although a few types are quite difficult. Heavy soldering of cast iron drainpipe is about the simplest. Sweat soldering of pipe joints is also fairly easy. Electronic soldering is the most difficult because joint condition is so critical.

This book covers the basics of gas and electric welder selection and the techniques of gas and arc welding, braze welding, brazing, and soldering. There is information on identifying and testing metals.

Safety is an important consideration. Working with high amperage electricity and very hot arcs or flames can be unsafe if certain basic precautions are not taken. These safety practices are discussed. If you pay attention to detail and follow directions carefully, you can handle welding tasks around the home or farm.

Chapter 1
An Overview of
Metal Joining Methods

Using heat to join metals provides certain advantages that no other methods can offer. Nuts, bolts, screws, and rivets can often be used more easily. Heat provides greater strength and vibration resistance in many cases. In other instances electrical (or heat) conduction is much greater.

Welding has developed into a series of very complex processes over the several thousand years since man first placed a couple of iron pieces in a hot fire and hammered the edges to create a joint. Welding classifications are now determined by heat intensity and heat source type. The equipment used may range from an oxyacetylene torch to a laser beam. For home and farm purposes, the cost and complexity of some of the equipment used in more elaborate heat sources are far beyond reason. This book focuses primarily on *oxyacetylene, oxy-MAPP*, and several types of electrical *arc welding*.

Welding is the fusion of similar metals by using heat to meld them. Forge welding probably started the historic processes. It can still be used, but is more suited to the work done by blacksmiths with wrought iron than to most modern repair and construction purposes. It's difficult to fit an I beam or plow with a 10-foot blade into a forge.

Gas welding offers the greatest portability at the most reasonable cost. It requires the mixing of oxygen with a

1

flammable gas such as acetylene, MAPP, or propane (oxygen itself is *not* flammable but supports the flame). Tanks in varying sizes are availale. Equipment can be portable as required or stationary and fed through manifolds. For home and farm purposes, manifold feeds are not needed.

Gases are fed through hoses from the tanks to the torch head, where they are mixed in the correct proportions to give the type of flame that does the best job. Torch types and tip sizes vary for metal types, thicknesses, and the jobs to be done.

Using electricity to weld is probably most often the choice of home welders. The equipment is relatively inexpensive. Essentially, electric arc welding passes a current through cables, one of which is clamped to the work surface as a ground. The other cable holds an electrode with which the welder strikes an arc. This melts the work surfaces and adds metal to the joint (Fig. 1-1).

Fusion welding can be used only for joining very similar metals. Stainless steel cannot be joined to mild steel in this manner any more than bronze can. When you want to join dissimilar metals, you need to do braze welding.

BRAZE WELDING

Braze welding is a process by which dissimilar metals such as mild steel and cast iron can be joined. You need enough heat to melt the braze rod, but not enough heat to quite melt the work surface metals. Braze welding doesn't offer quite as much strength as does fusion welding where the resulting joint is equal to and sometimes stronger than the joined metals. The braze welding process, though, requires less skill. It is a good starting point for the amateur craftsman. Bronze alloy filler rod is used for braze welding, along with a cleaning flux, to form the joint filler. Because less heat is needed for fusion welding, many people prefer it for work on metals like gray cast iron where the expansion and contraction of the base metals, through heating and cooling, may cause joint cracking during fusion welding. It often makes more sense to braze weld a cracked exhaust, intake manifold, or cylinder block on a car or tractor than to fusion weld.

Fig. 1-1. Arc welding in progress (courtesy of Lincoln Electric Company).

SILVER SOLDERING

Silver or *hard soldering* is tantamount to a breaking point in the use of heat for metal joining. Strength is lower than that of braze welding. Filler materials are different. The heat required is usually less. Silver soldering starts with a tighter joint design. The joint design should provide some joint support. This is not normally needed with fusion welding and braze welding. Silver solder is melted to the point where it will flow into the joint. At no time do the base metals being joined melt, though the filler material is appreciably stronger than lead/tin alloy solders used for soft soldering. For this reason, silver soldering (also called *brazing* at times) is used to make joints in electrical systems where vibration might crack soft solder. It is often used when assembling jewelry.

3

SOFT SOLDERING

While *soft soldering* isn't welding in any way, it is an extremely important metalworking process because it has many applications. The electrical conductivity of a soldered joint is difficult to surpass. Sweat soldered joints in pipe runs provide excellent and easy selling against the passage of liquids. Soldering involves melting a lead/tin alloy, and either settling a buildup of the alloy on the surfaces to be joined, or allowing capillary action to carry the alloy into the tight spaces of a joint, creating a seal. For electrical work, a soldering iron or gun is usually used. For sweat soldering pipe, a propane torch is usually needed.

HEAT SOURCES

Home and farm metalworkers are going to want less expensive equipment than professional welders will. Still, keeping the original outlay too low will sorely limit the jobs that can be done. Increasing equipment size may soon mean replacing the machinery at greater expense. Paying $500 for a setup that is seldom used to capacity is silly, but paying $150 for equipment and then having to hire someone to do the work is just as wasteful.

If all of your work is done with mild steel, and you do the work in the basement, garage, or outbuilding, your best choice is an ac arc welder of 300 or fewer amperes capacity (Fig. 1-2). The actual capacity needed depends on the thickness of the metal you are welding. Three hundred amps will be fine for most jobs you're likely to do, although some heavy farm welding might require more power. For very light welding uses, you can even purchase a 50-ampere unit to hold costs down. The unit available from Montgomery Ward can be plugged into small appliance 115-volt, 20-ampere circuits. You won't need to have 240-volt lines run into the building.

If you need portability and do most of your work with mild steel, the basic single stage oxyacetylene or MAPP gas torch units are excellent choices. Equipment costs will be higher than for arc welding units (though a lot less than portable arc welders). The units have the limitation of gas

Fig. 1-2. Portable arc welder (courtesy of Hobart Brothers Company).

welding equipment—almost totally useless on many nonferrous metals such as aluminum. Even with those metals, arc welders require special added extras to work well. In the past few years, gas welding has become more accessible to the amateur welder as smaller units have been introduced. The tools range from units produced by Cleanweld and Bernzomatic, suitable for work primarily with sheet metal and light tubing, to the oxy-MAPP torches made by Airco. Prices of the very small tools range from about $25 to $50. The larger and more powerful Airco unit costs nearly $200. Full-sized oxyacetylene torches and cylinders may cost more than $300, depending on quality, features, capacity, and tank size.

Flame features vary with the gases used. Oxyacetylene produces the hottest flame, around 5600 degrees Fahrenheit.

Oxy-MAPP produces a flame about 300 degrees lower. *Propane* provides the least heat of the available fuel gases—a temperature of about 4600 degrees Fahrenheit. Propane is also the most difficult of the fuel gases to control. It is hard to adjust away from an oxidizing flame, which can ruin ferrous metals being welded.

Once your gas flame is adjusted, you can start heating the work surfaces. When using an arc welder, make sure your grounding clamp is on the base metal to be welded quite securely, but far enough from the joint to be made to avoid harm. The welder is adjusted so the current suits the work type and electrode size. Then strike an arc with the electrode tip (make certain before striking that your face shield is down, for the arc is very bright and can cause eye damage) to get an arc temperature as high as 9000 degrees Fahrenheit. Working temperatures are more usually in the range of 6500 degrees Fahrenheit.

Until fairly recently, home and farm welders were limited in jobs. Most of the reasonably priced arc welders were capable of providing only ac to the work surface. Today, popularly priced ac-dc arc welders are readily found. Versatility has greatly increased. You are able to more or less custom design the weld for the job and the material. Negative (straight) polarity dc gives you greater heat on the work surface, which means you get more penetration with your filler material, along with less weld spattering. Positive polarity is used when you want less heat on the work surface (Fig. 1-3). This can be especially useful if you ever have to work with nickel or stainless steels. Note that ac arc welding

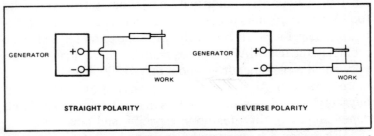

Fig. 1-3. Straight polarity and reverse polarity for dc arc welding (courtesy of Airco Welding Products, Murray Hill, NJ).

provides even heat because the polarity changes 60 times a second.

TIG WELDING

One of the adaptions possible with an ac-dc arc welder is the tungsten inert gas (TIG) welding process (Fig. 1-4). An inert gas such as *argon* is used to keep the weld more free of contamination. A weld with few oxide contaminants is stronger than a weld with many oxide contaminants. Shielding the weld to prevent oxidation makes for a stronger weld and often allows the use of smaller electrodes and less heat.

Although TIG welding equipment is not particularly cheap, the cost is now within reason for the serious home or farm metalworker; metal inert gas (MIG) welding equipment is not (Figs. 1-5 through 1-8). Metal inert gas welding takes over where TIG welding leaves off, at thicknesses over ¼ inch. The primary difference between the two processes is in the electrodes used. The TIG electrode is virtually indestructible tungsten, with filler metal added to make the weld. MIG electrodes are consumed as filler for the weld.

SAFETY

Safety considerations with any type of welding equipment are paramount. You are working with volatile gases or

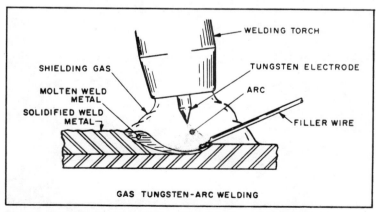

Fig. 1-4. Tungsten inert gas (TIG) welding (courtesy of Hobart Brothers Company).

with high amperage electricity and exceptionally high heat. If done correctly, welding can be as safe as almost any other home or farm workshop craft. If done incorrectly, injury or property destruction results.

Proper fusing and grounding of arc welders, along with using ground fault circuit interrupters, reduces problems. Concentrate on learning the basics of welding safety which are covered later in the book.

MAPP GAS TORCHES, PROPANE TANKS, AND SOLDERING GUNS

For hard soldering on small jobs, a handheld propane torch is usually more than sufficient, especially with the varied tip designs on the market. These tips do a more efficient job of mixing the fuel gas with air to provide the

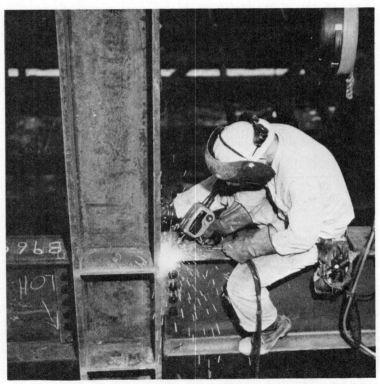

Fig. 1-5. A welder uses a Super MIG gun to join the structural section of a girder (courtesy of Airco Welding Products, Murray Hill, NJ).

Fig. 1-6. A welder is using a gun with an enclosed wire steel to join aluminum framework (courtesy of Airco Welding Products, Murray Hill, NJ).

temperatures needed in hard soldering and larger soft soldering jobs. Where greater amounts of heat need to be applied over larger areas, the MAPP gas torches are excellent. The MAPP/air mixture is several hundred degrees hotter than propane. MAPP gas, though, tends to be too hot for use in normal sweat soldering jobs. The job becomes more difficult as the extra heat makes the solder harder to control.

Whether welding or soldering, you should use the smallest possible heat source for the job, and then use the lowest possible setting. This will lessen metal warping and decrease tool cost. Consider that disposable one-pound propane tanks cost about one-third what MAPP one-pounders do. The waste is easy to see. Tanks can be changed, with one gas substituted for another in torches designed to use MAPP gas, but not the other way around.

The smallest heat sources used to join metals are the *soldering guns* and *irons* used to make electrical and electronic connections. Although their appearance is very different, their principles of operation are the same. Basically, the gun or iron is an electrical resistance device. The tip develops moderate amounts of heat. The iron or pencil shape

Fig. 1-7. A new gooseneck gun is used to join pipe (courtesy of Airco Welding Products, Murray Hill, NJ).

10

Fig. 1-8. A welding gun used to demonstrate the MIG process (courtesy of Airco Welding Products, Murray Hill, NJ).

is probably the one most often used for fine work. Some of the factory styles have very fancy and accurate temperature controls. Most have easily interchangable tips, and quite a few have different wattage heating elements, too. Soldering guns are gun-shaped tools with trigger controls. They usually have two heat ranges. The tips of wire heat nearly instantly. Guns usually offer more heat than do pencils (soldering irons, though, can get quite large and heavy). An average soldering gun will probably offer 125 to 200 watts of heating power. Soldering pencils will seldom exceed 50 watts. Gun tips can also be changed. The tip is applied to the work surface, and the solder is applied after the work surface warms up.

Chapter 2
Welding Safety

Safety is an essential ingredient to a good welding job. Sometimes you will be working with extremely flammable gases. Safety precautions are essential. In arc welding you will be working with extremely high amperage loadings. It is the amperage of electric current that does the most damage, not necessarily the voltage, so other types of safety needs must be met.

Both gas and arc welding produce extremely high heat. Precautions against fire must be taken. In addition, both methods of welding produce fumes. Some fumes are only irritating; others can be dangerous. Arc welding rays can cause painful skin burns. Eye damage may result if proper protective clothing and eye coverings are not worn. You must also guard against splattering hot metal.

ARC WELDING

Arc welding safety begins with equipment selection. Look for certification of the arc welding equipment by the National Electrical Manufacturers Association as to amperage and voltage ratings and for certification by Underwriters Laboratories.

Installation procedures for arc welding equipment follow more or less standard electrical wiring codes. Local codes, as well as the National Electrical Code, must be checked before circuitry is put in. First, you must know the size of the circuit breaker (or fuse) needed for your particular arc welder. For welders from 200- to 300-ampere capacity, you will need a 60-ampere circuit breaker. The circuit must be capable of carrying a 60-ampere load over the distance from the service panel entry. It is probably that for most runs of no more than 75 feet, you need at least number 6 wire. Longer runs may well require heavier materials. Don't skimp on such an installation. Do not try simply to pull a 30-ampere circuit breaker or fuse and add a heavier one for the same circuit, along with a new receptacle. The wiring is not heavy enough to stand the load. Your luckiest result will be a continual popping of the circuit breaker under moderate loads.

Most 30-ampere circuits will have, in newer installations, a number 10 wire. Circuits in older houses may have only number 12 wire if the run is under 50 feet. The maximum distance allowed with number 6 wire is about 95 feet, but I would prefer not to go past 75 feet without going to number 4 wire. The circuit must also be properly grounded to cut down on shock hazards. If you're working with older, ungrounded circuits, have the utility company install a suitable auxiliary panel.

Hazards

There are five basic hazards facing you when you get ready to arc weld: *electric shock, toxic fumes, arc radiation, flying sparks,* and *splattering metal.* Fire from either the sparks or the metal could be a hazard, but it is a result of other hazards and is secondary, though far from unimportant.

Electrical shock hazard is reduced by proper installation of the arc welder. It is further cut back by proper setup and handling of the arc welder during use. Assuming a correct installation and ground of your welder, any arc welder as ordered for home or farm use will have the correct size welding cables included. If your welder is a used one, replacement of the cables may be an immediate need. If so, cables up to 50 feet in length, for arc welders of up to 100 amperes capacity, should be number 4 in size. For welders of up to 200 amperes capacity, use a number 2 cable. You must select number 0 cables for the heaviest home welders, up to 300 amperes capacity. If for some reason you feel a need to increase cable length, then cable diameter must also increase. One hundred amperes from 50 to 100 feet will require number 2 cable, 200 amperes will need number 1 cable, and 300 amperes will call for 00 cable. Maintain the cables in good condition. Don't allow them to lie around where cars, lawn mowers, or anything else may be driven over them. Check your cables frequently for signs of wear like abrasion, cuts, and cracks. Replace damaged cables right away. Make sure all cable connections are tight before starting work.

Keep your work area clean and dry. A puddle of water underfoot is a hazard and a bad one. Trash lying around can also cause problems.

A properly ventilated work area is necessary. If ventilation is poor and can't be improved very much, use respiratory equipment. Exhaust fans are easily available and fit most window sizes. It might be a good idea to cut a hole in the wall for a fan.

Gloves

Gloves are important when you start arc welding. They protect your hands from splattering metal and high heat. Gloves help to make sure that a dry surface is handling the electrode holder. One type of glove I've used helps to keep hands even drier. Advance Glove produces a heat and spatter-resistant terry cloth glove that costs much less than a leather welding glove. It provides almost as much wear. Keep several pairs of these gloves on hand for those hot days when the gloves easily become soaked with sweat. Leather gloves provide more protection, but the terry cloth models are excellent, too. The gloves should have long gauntlet wrists to protect part of your forearms (Fig. 2-1).

In addition to long-wristed gloves of heavy material, you should always wear a long-sleeved shirt, with the sleeves down and buttoned, when welding. The shirt should be of cotton or wool. Synthetics melt too easily if a splatter hits them. I like the heavy denim cowboy shorts such as those made by DeeCee and Levi's because of the material's weight.

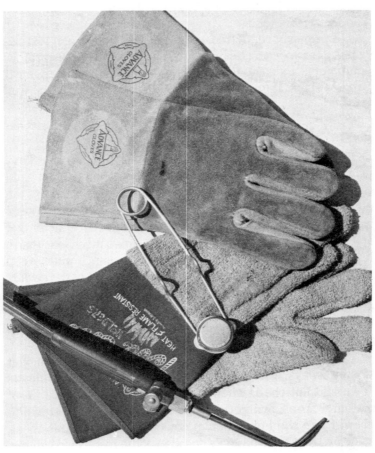

Fig. 2-1. Two types of welding gloves, both from Advance Glove Company. One type is made of leather, and the other type is terry cloth. Note the spark lighter and torch of Airco's Tote-Weld.

When you gas weld, long-sleeved shirts are a good idea, but they are essential for arc welders. Arc radiation can cause a severe skin burn, and metal splatter can cause even more severe burns. Arm protectors are available if you don't want to wear a heavy shirt in hot weather.

Eye Protection

Eye protection is as important as arm protection. First, the light from the arc is exceptionally bright when a strike is made. Even when the arc is maintained, that in itself can cause eye damage similar to what you might suffer staring directly at the sun. The *ultraviolet* radiation from the arc can cause damage if viewed from too close with no eye protection. Arc splatter can hit an eyeball. A dark shield is needed when working. The arc is so bright that you can't see what you're working on. While momentary exposure to the arc may do permanent eye damage, you can bet that any exposure to a piece of flying metal heated to 2000 degrees Fahrenheit will be harmful.

Eye protection for the arc welder consists of a face-covering shield, with a standard size (2 by 4¼ inches) glass window through which you can view the progress of your work. Various densities of filter material are available. The type of work determines the density, so you'll eventually need extra filters. Most home and farm arc welding can be safely accomplished with a number 10 filter in place in the shield. Things change when you start working with TIG or MIG equipment, though. If you are welding at 200 amperes, TIG style, you will need a number 14 filter. A number 12 filter is required for operations under 200 amperes. If you decide to use a carbon arc torch to cut metal, then you also need a number 14 filter.

The first few times you slip the shield over your head, you'll almost certainly experience claustrophobic feelings. It is dark, even with a number 10 lens (Fig. 2-2). Striking an arc after dropping the shield in place takes a little practice. That arc, once struck, will provide you with sufficient light to carry on with the weld or the cut you're making.

If at all possible, when selecting your shield, choose one

with a chemically treated cover glass. A cover glass of this type has a life about five times longer than standard welding glasses. It is well worth the modest extra cost.

Footwear

Shoes or boots should be sturdy and in good condition. It is never a good idea to wear lightweight shoes like sneakers or running shoes when welding or cutting. Splattering metal and slag from the cut can burn through these shoes instantly. If your shoes are rubber or plastic, even more damage can be done to your feet when the material melts. Shoes with nonskid soles should be worn.

FIRE PREVENTION

Fire prevention is an important part of welding. It starts with having the right kind of fire *extinguisher* on hand. Move all combustible materials out of the welding or cutting area. If a fire starts, keep it small and put it out rapidly.

Always keep at least one moderate sized dry chemical fire extinguisher on hand. Fire extinguishers come in many sizes. There are extinguishers for each of the four fire classes. Those meant to be used on wood and paper fires (*class A*) are usually soda/acid types. These extinguishers use water to put out fires. These extinguishers are generally not good for many of the fires that may start around welding equipment. They are dangerous when used around electrical fixtures of any kind. There is one time, though, when such extinguishers, or a pail or two of water, should be kept on hand. Once I had to do some cutting and welding on a disc harrow that broke in a field of dried hay. The cutting caused the hay to catch fire, and water was needed to put it out.

Class B fires involve flammable liquids such as oil, gasoline, kerosene, oil-base paints, varnishes, and other similar materials. The vapor on or just above the surface of the material burns, and you need to cut off the supply of oxygen. Water will only spread the flames and make things worse. In almost all cases the flammable liquids are lighter than water. The liquids will float to the surface and spread as the water does. Foam extinguishers can be used, but most often class B

Fig. 2-2. Helmet and lens for electric arc welding (courtesy of Hobart Brothers Company).

extinguishers are of the carbon dioxide type. This heavy gas is also very cool. It helps to cut the temperature of the fire below flash point, as well as cutting off the oxygen supply. Sand is also effective on smaller fires where you can get close without being in danger.

Class C fires are electrical in nature and are fought in the same manner as class B fires. Use a dry chemical, or perhaps a foam to cut off the oxygen supply. Under no circumstances should water ever be used to put out an electrical fire.

Class D fires are of little interest to the home and farm welder. These fires are caused by burning magnesium. Magnesium is seldom used in areas where a homeowner or farmer will be welding. You must have a special fire extinguisher to put out a Class D fire as magnesium forms its own oxygen, unlike most flammable materials, and will continue to burn if a class A, B, or C extinguisher is used.

Fire extinguishers are labeled as to the type of fire they are most effective against. Class A fires are the only type that may be extinguished with water. Class B fires require a smothering action, so sand or chemicals may be used. Class C fires are best extinguished with chemicals. Fire extinguishers that don't use water will be suitable for both class B and C fires, and a few are also rated for class A fires. Generally, those extinguishing materials most useful for class B and D fires lose effectiveness when used on class A fires. When working with cast iron, the only type of fire extinguisher you should use is a dry chemical type. Both water and carbon dioxide drop the temperature of the metal so rapidly that it is likely to crack or even shatter. If a fire starts at a piece of metal being welded or cut, it is again best to use a dry chemical extinguisher to prevent severe metal warping.

Make sure your extinguisher is small enough to handle easily but large enough to put out a fire of moderate size. Some can be refilled. I prefer the slightly larger models that can be refilled, but at times getting a refill can prove a hassle. The companies doing the recharging don't want to bother with one small unit at a time. All fire extinguishers should be checked for charging at regular intervals.

Check to make sure there are no explosive vapors in the air. A can partly filled with gas for the lawn mower sitting 10 feet out on the lawn may not seem like much of a hazard, but there is no sense taking a chance on that one random spark. Move lawn mowers, chain saws and other gasoline-powered tools out of the way. Heavy dust buildup can also be a problem. Some dust becomes explosive when dispersed in the air at the right concentration. Allow dust from sweeping to settle before starting work.

Various companies can supply you with flameproof curtains of asbestos or other fire-resistant materials for use as welding curtains. Such curtains are exceptionally handy when you have to work inside a frame building or around combustible materials or objects that cannot be moved. If the floor is of combustible material, it also must be protected.

If metal or asbestos is not available for floor protection, use an inch or more of damp sand. The use of any damp material with electric arc welding equipment requires special precautions against shock. Wear rubberized, lined welding gloves and heavy rubber-soled shoes with no nails in the soles. Keep a careful check on system grounding.

You must protect the welding equipment from any moisture. Also, look carefully at the surface on which you are welding. Never cut or weld on ordinary stone or brick. The water in the molecules will expand from the extreme heat generated by the torch or arc. This may cause the stone or brick to explode. Use only asbestos padding or firebricks to make welding workbench surfaces, and place one or the other under any welding being done on concrete floors.

Additional safety precautions involve common sense. When cutting metal, make the cut away from yourself. When doing overhead welding or cutting, keep the electrode well away from your body. Make sure you wear a heavy arm protector that goes well up your shoulder. Don't let the electrode holder just lie on a grounded metallic surface. Allow only qualified repairmen to work on the interior of your arc welder. Don't place the welder on a dirt floor, no matter how dry it seems. Don't ever use defective equipment, whether it is an electrode holder with broken or weak jaws or cables that are cracked or badly abraded.

GAS WELDING

Although many rules for gas welding safety are similar to those for arc welding safety, there are some major differences. You seldom have to face the problems of electrical shock hazards, but you will face the same difficulties with toxic fumes, bright light (though not as bright as the arc), combustible working surfaces, spatter, and slag. Other problems include overheating high pressure gas containers, the possibility of incorrect hose hookups, the chance of cutting the gas hoses with the torch flame, and so on. In place of the arc welding cables, you are faced with hoses that can be cut, cracked, or abraded. Instead of an electrode holder or carbon arc torch, you have a welding torch and a cutting torch which may suffer in the same ways.

Safe Cylinder Handling

You must keep oxygen, acetylene, MAPP, and propane gas cylinders in a spot protected from flying sparks, open flame, and hot slag. Cylinders must never be abused. Cylinders must be either mounted securely in a stationary position or secured in a stable cart to keep them from falling. Cylinder connections are even more sensitive to abuse than the cylinders themselves. Never force any fitting that doesn't seem to wish to go on, and never tighten any connections with a pair of pliers. The metal used is usually brass, which is relatively soft and mars easily. Use either the proper size wrench or a top quality adjustable wrench to make the connections. Don't use wrenches to open tank valves.

Regulator valves must always be attached to high pressure cylinders before attaching the oxygen equipment. Acetylene and other fuel gas cylinders should have regulators attached prior to the time you attach hoses for use.

Avoid any compound or pipe fitting that might contain petroleum or its by-products. If lubrication is needed, stick with *glycerine* or a lubricant specifically recommended for oxygen equipment. Your equipment's fittings should never require lubrication.

Hose Care

Your first job with the hoses is to make certain you have the proper sizes and types for the uses to which you'll be putting the equipment. The oxygen hose is nearly always green. The acetylene, or fuel gas, hose will be red. Threads on the oxygen hoses will be right-hand style. The fittings on your fuel gas hose will have left-hand threads. The acetylene connection nuts will have a groove around them to help you make the identification.

A new hose should be blown out before the first hookup, because there is talcum powder inside it. Fine dust can clog your torch. Both oxygen and acetylene hoses can be blown out with oxygen (never a fuel gas for both hoses), but you'll find it safer to use filtered, (for water) compressed air in the acetylene hose. Test all fresh connections for leaks before shutting down the torch valves and opening the pressure regulator at the tank to its normal working pressure. The test requires you to shut down the torch valves firmly, then open the oxygen pressure regulator to allow pressure to flow. Adjust to 10 pounds per square inch. Linde recommends the use of soapy water to make the test. If leaks do occur, shut down the tank valves, bleed the hoses by opening the torch valves, and back off the connections. Remove and tighten the connections as the manufacturer directs. If troubles persist, have your distributor check the connections.

Use the shortest hose lengths possible. This helps prevent kinking and possible premature hose damage and wear. Then keep the hose from underfoot so it is not driven over or slipped through oil and grease. Protect the hose from hot slag, sparks, spatter, hot metal, and an open flame, just as you do the cylinders.

A periodic check of your hoses for wear and deterioration is an extremely good idea. Leaks that occur around the areas where fittings are attached to the hose can usually be rapidly repaired by cutting the hose at least a couple of inches back of the worn area and replacing the fittings. After two months of heavy use, the hose should be trimmed back at least 2 inches. The fittings should be replaced or reinstalled. If deterioration or damage along much of the hose is great, then it's best to replace the complete unit—hose, fittings, and all.

Flashback and Backfire

Two specialized problems exist with all oxygen welding and cutting equipment, from the smallest Bernzomatic and Solidox torches to the largest and best commercial style units. *Flashback* is the more dangerous of the two problems. The flame actually jumps back into the torch and may move back to reach the hose and regulators. If you hear the characteristic shrill hissing sound, immediately cut off the oxygen at the torch. Cut the acetylene at the torch, and cut off both the oxygen and acetylene at the regulators. Make no attempt to relight the torch until it has plenty of time to cool down. With flashback, it is best to examine the torch and consider the operation of it before any attempt to relight it is made. A clogged *orifice* and incorrect oxygen and acetylene pressures are usually the causes of flashback. Keep gas pressures down where they belong, and make sure all torch tip orifices are clean. After you check the tip and the pressures, clear the torch by opening the oxygen regulator valve and the torch valve for a few seconds. This will blow out any soot which may have accumulated in the torch during flashback.

Backfire is a simpler and not quite so dangerous phenomenon. The torch flame goes out with a loud snap or pop. First, shut off the torch valves. Make a check of all connections. Before relighting, it's a good idea to check the tip and the head of the torch for looseness. Also, check for dirt on the valve seats. If none of these problems exist, then it's likely you somehow touched the tip of the torch to the work in progress, causing the backfire.

Some further safety precautions with oxygen equipment involve lighting the torch. Never use matches; use a spark lighter instead. Matches held in front of a torch provide an excellent way of finding out just how painful burns can be. The spark lighter allows you to keep your hand several inches back from the flow of gases. Make sure you're already wearing the glove on your torch-lighting hand. Torches that go out should never be relighted from a hot spot on the work surface.

Eye Protection

Because gas welding, cutting, and brazing don't give off ultraviolet radiation as arc welding does, a bit less eye protection is needed. Too many people, though, go without any eye protection. Goggles are important (Fig. 2-3). Different shades or densities of lenses are available for different jobs. Gas brazing requires a number 3 shade lens in the goggles, but you may wish to go to a darker number 4 if the work is very heavy. As work heat increases, so does the need for a darker lens shade. Basically, welding metals to ⅛ inch requires at least a number 4 lens. From ⅛ to ½ inch, your best protection is at least a number 6 lens. For more than ½-inch thicknesses, you may need the protection provided by a number 8 lens, though most of the time a number 7 will do the job.

Oxygen cutting requires good eye protection also, with cuts of metal less than 1 inch thick needing at least a number 3 lens. A number 4 lens is preferred. Cuts up to 6 inches require a number 4 or 5 lens.

Don't even bother trying to depend on eyeglasses or sunglasses for protection while gas welding. When excess heat is present, eyeglasses tend to slip toward the tip of the nose. There is not a bit of protection around the sides of the eyeglasses, but a good pair of welding goggles has heat-

Fig. 2-3. Goggles for gas welding (courtesy of Airco Welding Products, Murray Hill, NJ).

resistant material to keep hot metal from coming around the edges and blinding you.

As with arc welding, goggles of the proper lens shade should always be provided for spectators and equipment operators. Often, for both operators and spectators to get a good view of the welding process, it is necessary to get the face relatively close to the work. Still, keep as far away from the weld as possible.

If round goggles do not fit well over your glasses, obtain *square front goggles.* These will fit easily over almost any style of eyeglasses.

Chapter 3
Metal Identification
and Rod Selection

The wide variety of welding rods and electrodes available can at first seem intimidating, but some knowledge about metals, and the rods or electrodes, makes the actual selection fairly simple. The first step in any welding job is to find out just what metal you are dealing with. You can pick a rod or electrode of the correct composition and size for the type of metal and its thickness. Pay attention to the joint strength needed. Joint strength is influenced not only by the rod selection, but also by the type of joint to be made.

Although a basic knowledge of metallurgy is not really essential to most welding—probably 98 percent or more of farm and home welding will be done on mild steel, cast iron, or aluminum—a look at some tests to determine what kind of alloy you are working on can be helpful in making a neat, long lasting joint. Some metals require specialized equipment and techniques that you may not wish to get involved with because of cost or complexity. There is certainly little point in spending hundreds of dollars to get TIG welding equipment if it will be used only once.

TESTING METALS

There are several ways of determining the type of metal involved on a job; most are the same tests used by professionals in the field. Every type of metal test generally used in

metal / test	low carbon steel	medium carbon steel	high carbon steel	high sulphur steel
appearance	DARK GREY	DARK GREY	DARK GREY	DARK GREY
magnetic	STRONGLY MAGNETIC	STRONGLY MAGNETIC	STRONGLY MAGNETIC	STRONGLY MAGNETIC
chisel	CONTINUOUS CHIP SMOOTH EDGES CHIPS EASILY	CONTINUOUS CHIP SMOOTH EDGES CHIPS EASILY	HARD TO CHIP CAN BE CONTINUOUS	CONTINUOUS CHIP SMOOTH EDGES CHIPS EASILY
fracture	BRIGHT GREY	VERY LIGHT GREY	VERY LIGHT GREY	BRIGHT GREY FINE GRAIN
flame	MELTS FAST BECOMES BRIGHT RED BEFORE MELTING	MELTS FAST BECOMES BRIGHT RED BEFORE MELTING	MELTS FAST BECOMES BRIGHT RED BEFORE MELTING	MELTS FAST BECOMES BRIGHT RED BEFORE MELTING
Spark*	Long Yellow Carrier Lines (Approx. .20% carbon or below)	Yellow Lines Sprigs Very Plain Now (Approx. .20% to .45% carbon)	Yellow Lines Bright Burst Very Clear Numerous Star Burst (Approx. .45% carbon and above)	Swelling Carrier Lines Cigar Shape

*For best results, use at least 5,000 surface feet per minute on grinding equipment. (Cir. x R.P.M.) / 12 = S.F. per Min.)

metal \ test	manganese steel	stainless steel	cast iron	wrought iron
appearance	DULL CAST SURFACE	BRIGHT, SILVERY SMOOTH	DULL GREY EVIDENCE OF SAND MOLD	LIGHT GREY SMOOTH
magnetic	NON MAGNETIC	DEPENDS ON EXACT ANALYSIS	STRONGLY MAGNETIC	STRONGLY MAGNETIC
chisel	EXTREMELY HARD TO CHISEL	CONTINUOUS CHIP SMOOTH BRIGHT COLOR	SMALL CHIPS ABOUT 1/8 in. NOT EASY TO CHIP, BRITTLE	CONTINUOUS CHIP SMOOTH EDGES SOFT AND EASILY CUT AND CHIPPED
fracture	COARSE GRAINED	DEPENDS ON TYPE BRIGHT	BRITTLE	BRIGHT GREY FIBROUS APPEARANCE
flame	MELTS FAST BECOMES BRIGHT RED BEFORE MELTING	MELTS FAST BECOMES BRIGHT RED BEFORE MELTING	MELTS SLOWLY BECOMES DULL RED BEFORE MELTING	MELTS FAST BECOMES BRIGHT RED BEFORE MELTING
Spark*	Bright White Fan-Shaped Burst	1. Nickel-Black Shape close to wheel. 2. Moly-Short Arrow Shape Tongue (only). 3. Vanadium-Long Spearpoint Tongue (only).	Red Carrier Lines (Very little carbon exists)	Long Straw Color Lines (Practically free of bursts or sprigs)

*For best results, use at least 5,000 surface feet per minute on grinding equipment. (Cir. x R.P.M. 12 = S.F. per Min.)

Fig. 3-1. Metal testing chart (courtesy of Hobart Brothers Company).

industrial welding, within reason, can be adapted to use around the home and farm.

Visual Test

Your first and simplest test is simply to look at the metal. Figure 3-1 indicates the *visual* properties of metals likely to be found around homes and farms. Most people with near normal eyesight can tell most of the ferrous metals from nonferrous metals. Brass looks nothing like cast iron. Many of the ferrous metals provide their own identification problems. Mild steel is medium gray along a fracture line, so it is easily identified if you're comparing it to the bright gray fracture line of low carbon steel. Making the identification from an unfinished surface is often extremely difficult. Both steels have dark gray surfaces, often with visible machining or forging marks. Usually low carbon steel is cast. If mold marks are not machined off, they can help identify it. If the two metals are freshly machined, visual identification cannot be made. Both are then very smooth and a dark gray. Thus, simply looking at a metal and comparing it to a chart has severe limitations, especially if there isn't enough of the material available to you for a fracture test.

Chip Test

Chip testing requires that less material be wasted than does fracture testing. Chip testing will give reasonably accurate results in most cases. Chip testing makes it easy to identify low, medium, and high carbon steels. The lower the carbon content, the more easily the cut can be made and, usually, the easier it is to run a continuous strip or chip. You use a handheld cold chisel to get the chips. Steels that are difficult for you to chip have a high carbon content. Steels that can't be chip tested are cast steels. These require spark testing to determine their composition. The reasoning behind using the spark test for these two steels is quite simple; any chips you do manage to get from cast steel and wrought iron will appear to be just about identical.

Spark Test

For *spark tests,* which are more difficult to read results from than are the preceding two tests, you'll have to have a good quality grinding wheel and a chart of simple spark patterns. When the metal is held against the grinding wheel, you can analyze the spark stream flying off the wheel as it tears off hot metal particles. Practice is needed to make it easy for you to determine just what spark pattern you actually have. While some spark patterns are similar to each other, each metallic alloy does give off a spark stream with a different pattern or color. In the spark pattern you'll find that the spark stream length will vary from 70 inches for machine steel down to none for copper or brass (the latter two metals are useful in making tools for use around areas conducive to fires or explosion).

Stream color will vary depending on the stream's distance from the wheel. Wrought iron produces a straw-colored stream close to the grinding wheel, but the sparks at the end of the stream will be white. Type 410 stainless steel gives the same basic coloring as wrought iron, straw to white, but it produces a shorter spark stream with a moderate number of spurts. Wrought iron provides few spurts. Also, the volume of the wrought iron stream is larger than the stream from the stainless steel.

Spark stream analysis becomes easier if you find the various parts of the stream (Fig. 3-2). It begins with wheel sparks and develops a center stream section followed with the tail section. For best results, obtain some previously identified samples of as many of the metals you expect to be working with as possible. Then practice spark testing and analysis. Use marked and unmarked samples of each metal. Place all metals to be tested in a box. Take out an identified piece, and then compare it with unidentified pieces until you get a match. You can simply keep your marked samples on hand and compare those with the metal on any particular welding job until you find a match. This takes longer if you do much welding, but it works.

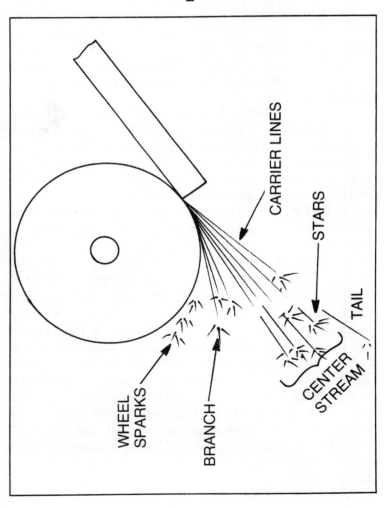

Fig. 3-2. Spark stream parts.

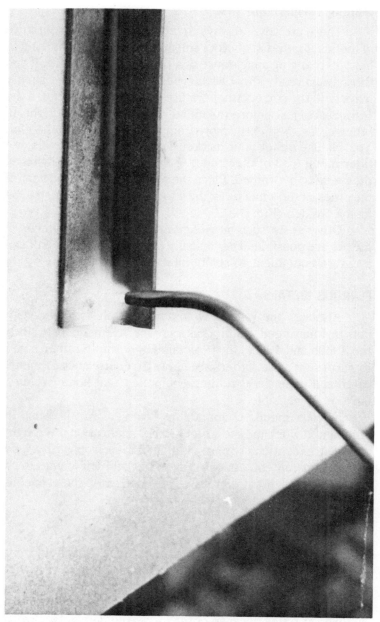

Fig. 3-3. When testing metal, if the rod and metal of about the same thickness melt at the same temperature, the alloys are similar enough for fusion welding.

Melting Temperature Test

There are several ways to test the melting temperature of metals. Many distributors sell marking pencils that react to heat by changing appearance at a specific temperature. If you think you already know what type of metal you have, simply mark with the correct temperature pencil for that metal. If the mark reacts just before the metal melts, then you're right. In other cases, a slightly more complex procedure is helpful. You can use either your marked series of sample metals or a filler metal (rod or electrode) of the type you think matches the metal to be welded. Place the pieces together and heat. If they melt at the same time, the alloys are similar enough to be fusion welded (Fig. 3-3).

Other tests such as weighing similarly sized pieces of metals are possible. They usually require too much skill and accurate equipment to really interest the amateur.

FERROUS METALS

Although the variety of common metals found in products that may need welding has increased tremendously since the 1940s and 1950s, most metals fall within a rather small range of similarity. Others are likely to require such sophisticated and expensive welding equipment that home welding isn't really sensible.

Ferrous metals contain iron. Wrought iron is found in ornamental work and a few other minor applications. Wrought iron is an exceptionally pure version of basic iron, with a very low carbon content. At one time industrial and commercial application was great, but it was reduced when the economical production of steel became possible.

Cast Irons

Cast iron can be classified as one of four types: *gray iron, malleable iron, pig iron,* and *white iron.* Various alloys with carbon, silicon, and other elements, along with heat treating, are used to produce cast irons. There has been a shortage of cast iron at times because of the increased demand for wood stoves. Gray cast iron can be welded with either oxy-

34

acetylene or arc welding equipment if certain precautions are taken. Cast irons have a carbon content of from 1.7 percent to about 4.5 percent. They are not malleable (except for the special category of malleable iron). Gray cast iron is found in engine blocks, wood stoves, and fire hydrants. The composition of gray cast iron is usually 1.7 to 4.0 percent carbon, a minimum of 1 percent silicon, .2 percent sulfur, three times that amount of phosphorus, and no more than 1 percent manganese.

White cast iron is used for plows and a variety of other heavy-wear surfaces. It shows silvery white along its fracture lines. White cast iron is less widely used than gray cast iron. It is difficult to machine, very hard, and brittle. Composition and rapidly cooling molds give white cast iron its exceptional hardness. The chemical structure is similar to that of gray cast iron with 2.5 to 4 percent carbon, 0.4 to 1.6 percent silicon, about 0.15 percent of sulfur, 0.4 percent of phosphorus, and from 0.3 to 0.8 percent of manganese.

Malleable cast iron usually has a slightly lower carbon content than other cast irons, from 2.0 to no more than 3.0 percent, while the silicon content rises from 0.9 to 1.8 percent. Both sulfur and phosphorus contents must not exceed 0.2 percent, and manganese content will range from 0.25 to 1.25 percent. Although these differences may seem minor, they account for substantially different properties in the alloy when heat is added. Malleable cast iron is used in small tools, farm implements, and machine parts because of its great strength and *ductility,* overall toughness, and resistance to shock.

Pig iron comes to the foundry to be used in the manufacture of other irons and steels. Other iron alloys are around, but they are of little use to the do-it-yourself welder.

Cast iron requires several special welding techniques, from preheating—whenever possible, as preheating reduces shrinkage and further cracking of the base metal—to special fluxes to keep too many oxides from forming. Fortunately, the flux problem is often taken care of by the rod or electrode manufacturer, while preheating of some smaller parts may not be essential. Even larger parts, if they are not subjected to shock loadings, don't always need to be preheated. Thorough preheating is often difficult or impossible because of the part's size.

Such electrodes and rods as Hobart's *Nickelcast,* Airco's *No. 9 sintered cast iron,* and Eutectic's *Xyron 2-24* can be used for almost any cast iron welding job. Nickelcast electrodes can weld cast irons together or to some dissimilar metals. Dissimilar metals can be welded together, usually with a mild steel being welded to some form of cast iron. The joint will be good, though not nearly as strong as if done only in mild steel. With the electrode made of about 90 percent nickel, it is obviously a high nickel content filler for general use and is easily machinable. Airco's No. 9 welding rod is primarily recommended for gray cast iron. You must use a separate flux for sound welds. Eutectic's Xyron 2-24 electrodes are specifically designed for use with water jacketed engine blocks and provide minimal surface penetration where those casting walls are thin.

Alloys Used in Steel

Steel production is more complex than is iron production. The alloys produced need to be purer. They have many different properties.

☐ **Chromium:** improves hardness (used in small amounts) and is used in stainless steels and tools.

☐ **Cobalt:** aids cutting ability and is used in high speed cutting tools.

☐ **Manganese:** increases abrasion resistance and toughness; used in rails.

☐ **Molybdenum:** adds to ductility, strength, and shock resistance; used in machine parts and tools.

☐ **Nickel:** resists heat and adds strength when used in large amounts; used in stainless steels and corrosion-resistant tools.

☐ **Silicon:** increases strength; and is used in precision castings.

☐ **Sulfur:** aids machinability and is used in machine parts.

☐ **Tungsten:** helps steel retain its hardness and toughness at high temperatures; used in magnets and high speed cutting tools.

☐ **Vanadium:** aids in increasing steel strength and ductility; used in springs, tools, and some machine parts.

☐ **Zirconium:** provides a fine grain; used in machine parts and tools.

The alloying processes for steel can use several elements at the same time. Many modern motorcycle frames (usually on the more expensive off-road models) are made of *chrome-moly steel.* The major alloying elements are chromium and molybdenum. One increases the hardness of the alloy and the other increases its shock resistance and strength. This steel is not mild and requires specialized welding equipment and plenty of skill to work properly. Either TIG or MIG welding is used, depending on the steel's thickness.

Mild and Low Alloy Steels

Mild steel will have only a small percentage of carbon, with only tiny amounts of alloys to add to hardness. Mild steel is used in car bodies and frames, motorcycle and bicycle frames, truck and bus bodies, so-called wrought iron lawn furniture, ships and larger boats, and many types of wood stoves. Mild steel can be bought in many forms, is easily worked, and is easily welded in almost all its forms. Whether it is pipe, sheet or plate, or rod stock, the proper rod or electrode will nearly always allow you to produce a strong, attractive weld bead in mild steel.

Mild steel welds might have to be made in almost any position imaginable—from flat, vertical, and overhead—and most companies have developed fast freeze electrodes. Lincoln's basic. Fleetweld 5 fast freeze electrodes is a good example. It is used for all general mild steel applications, both for fabrication and maintenance welding. The electrode provides deep penetration, works well on galvanized, plated, or grimy steels, and can also be used on sheet metals. The Fleetweld 5 matches the *American Welding Society's* (AWS) specifications for an E6010 electrode and can be used with dc welders. For ac welders, Lincoln produces its Fleetweld 35,

AWS E6011. Airco's No. 7 mild steel rod does the general purpose job for gas welders. Hobart has a version of the AWS E6010 electrode cataloged as Hobart 10, with their 335A matching the ac requirements of AWS E6011. While Eutectic doesn't state which of their rods and electrodes conform to AWS specifications, the correct mile steel electrode would be the *Steel-Tectic,* according to Mick McGarry, technical director.

It is difficult to separate low alloy steels from carbon alloy or mild steels. Most carbon steels are alloyed with other substances such as silicon and manganese. If the manganese content of the steel alloy is below 1.7 percent, then the steel is a simple carbon or mild steel alloy. Once the manganese rises to about 1.7 percent, the steel offers greater strength and becomes a low alloy steel. It will often be more difficult to weld than mild steel. A metal test is needed so that you can select the correct rod or electrode. You can no longer simply rely on an AWS E6010 or E6011 electrode to do the job well. If the alloy changes much more, you may have to invest in a tungsten inert gas addition for your arc welder to do maximum strength welding. Alloy steels are used only where maximum strength is essential. Difficulty in welding alloy steels is often due to weld contamination by atmospheric elements. Even the shielded arc process doesn't help until extra protection is added to keep air from reaching the filler metal and base metals as they are puddled. The air reaching the molten metal forms oxides and other impurities, which decrease weld strength. In noncritical applications many alloy steels may be arc or gas welded with minor adjustments and no extra equipment.

Low alloy steels are often listed as mild steels. The increase in welding difficulty is handled by simply selecting the proper rod or electrode. A useful electrode on mild and alloy steels is Lincoln's *Jetweld 1* (AWS E7024), which provides a tensile strength of 72,000 to 91,000 pounds per square inch (psi). Lincoln's *Fleetweld 5* has a tensile strength of 62,000 to 69,000 psi. Hobart's 24H produces tensile strengths up to 83,500 or 84,000 psi versus the Hobart 10's 62,000 to 76,500 psi. Electrode composition is different, too. Hobart's

10 has .07 percent carbon, .5 percent manganese, .25 percent silicon, .016 percent phosphorus, and .02 percent sulfur. The 24H contains .084 percent carbon, .76 percent manganese, .61 percent silicon, .018 percent phosphorus, and .010 percent sulfur. The higher manganese and silicon contents are extremely important to the strength qualities of the final weld.

Chromium and Nickel Chrome Steels

Chromium and *nickel chrome steels* are basically *stainless steels.* These steels are more difficult to weld properly than are mild and low alloy steels. One type of straight chromium stainless steel, *ferritic,* will become brittle when welded and needs to be *annealed,* a process that requires reheating to just below a critical temperature. A second type of chromium stainless steel is *martensitic.* This steel can be heat treated back to hardness after welding, but it must be preheated and often postheated to prevent cracking. The use of small diameter electrodes is also recommended. You should use a series of passes to build up the filler metal and not try to do the job in a single pass. This keeps down excessive heat buildup to a degree.

Chrome or nickel chromium stainless steels are heat treatable and the easiest of the stainless steels to weld. A common problem is corrosion between the grains of the metal brought on by a switch in positions of the carbides during welding or heat treating. Annealing will cure the corrosion problem, along with special alloys designed to prevent it. Such alloys use extremely low amounts of carbon—no more than 0.03 percent—and do not produce corrosion. Stabilized stainless steel also gets around the problem by alloying in either *titanium* or *columbium.* These metals combine more easily with carbon than does chromium, thus preventing the position shift of the carbides when heat is applied.

Stainless steel electrodes of various types are readily available. Hobart's 308-15 is designed for low alloy types of stainless steel, while their Type 308L is suitable for ELC *(extra low carbon)* stainless steel. Hobart produces an electrode for use with nearly every type of stainless steel made.

Lincoln offers similar types, and Airco has about 18 different stainless steel electrodes. Eutectic Corporation also offers several stainless steel electrodes, including one intended for use with stainless steels of unknown types. In almost every instance, weld production will be better if you can find out the number designation of the stainless steel and match the electrode to the base metal. Where the stainless steel type remains unknown, you'll find it best to anneal the resulting joint. Heat treating is not always practical; annealing makes more sense in many cases.

NONFERROUS METALS

The use of *nonferrous* metals in various applications is quite widespread today—copper pipe and wire; aluminum frames for chairs, car bodies, and other things; brass or bronze fittings and tools; and occasional bits of magnesium for super lightweight applications. Magnesium is seldom used in nonindustrial applications. The welding processes are very similar to those used for aluminum and its alloys, with one exception. If you ever have to weld magnesium, do not do so unless you have a special class D fire extinguisher to put out possible magnesium fires.

Aluminum

Aluminum can be welded with oxyacetylene equipment or with manual shielded arc equipment. When exposed to air, aluminum oxidizes quickly so that extra shielding often results in much better welds. A metal inert gas or tungsten inert gas arc welding setup will usually do the best job. You should remember two other things when welding aluminum. The base metal becomes extremely weak as it reaches melting point, thus forcing you to make sure a solid support prevents collapse of the pieces being welded. There is no change in the metal color as the melting point is approached, and the metal appears to just collapse in an instant (ferrous metals all change color extensively as their melting points are approached).

Many aluminum alloys are available including commercially pure aluminum (aluminum that is 99 percent pure).

Wrought alloys of aluminum contain many of the same elements as do alloys of iron: silicon, manganese, chromium, and nickel. You can also look for aluminum alloyed with copper, magnesium, zinc, zirconium, vanadium, lead, and bismuth. Aluminum is found in numerous forms and shapes.

Eutectic has a brazing rod (*Eutec. Rod 21FC*) with a melting point below that of the base metal. This rod can be used with furniture joints and gives you a surprisingly high tensile strength. It is also factory fluxed. For heavier applications, Eutectic offers *EutecRod 2101*. Hobart produces only solid aluminum welding wire for use with TIG or MIG welders, while Lincoln has the *Aluminweld A1-43*, a flux-coated electrode. Airco offers Nos. 43 and 57 rods, along with phosphor bronze and aluminum bronze electrodes used in brazing aluminum.

Copper Alloys

Copper or one of its alloys may need to have the strength associated with fusion welding. Copper is a metal that has almost literally no plastic range; that is, it remains totally solid until the metal becomes liquid. Like aluminum, copper has a high rate of heat conductivity and expands a lot during heating. It contracts equally as much as it cools. The great heat conductivity means that it takes longer to produce a molten puddle with which to work (covering the base metal close to the area being welded with a heat-resistant material such as asbestos can be a help). Generally deoxidized copper is the only type that should be welded where joint strength is important.

Electrolytic copper infuses the welds with oxides from the base metal, thus causing weak joints and forcing a strength reduction that can reach as high as 60 percent of normal joint strength. If electrolytic copper is used for welding purposes, the use should be reserved to ornamental or other applications where joint strength is not critical. In such cases, use hard soldering or braze welding techniques. In instances where welded joint strength is needed with copper, you can use EutecRod 1804 or Airco's No. 70 phosphor bronze rod.

For high ductility and great strength, *cupronickel* consists of 30 percent nickel and 70 percent copper. Nickel silver contains about 66 percent copper, with the remainder of the alloy split fairly evenly between zinc and nickel. *Monel* is another copper nickel alloy, and it is highly resistant to acids. It contains about 28 percent copper and 67 percent nickel. The remainder of the monel alloy consists of iron and manganese. Monel may also be alloyed with as much as 2.75 percent aluminum. Hardness and strength is about equal to heat-treated steel alloys.

Inconel is an amazingly strong metal, with a possible tensile strength ranging up to about 185,000 psi. Inconel consists of chromium, nickel, iron, and a few other alloying agents added to copper.

Brass and Bronze

Brass is an alloy of copper and zinc, with possible variations including the addition of lead or aluminum. Usually braze welding and hard soldering are more satisfactory than fusion welding on these materials, though welding rod and electrodes are available if the extra strength is needed.

Bronze is basically an alloy of tin and copper, with greater strength and ductility than brass. Lead is a normal third part of the alloy, as are small percentages of zinc and nickel. *Phosphorus* can be added to deoxidize the alloy, which results in an alloy called *phosphor bronze*. Aluminum bronze has higher corrosion resistance when manganese, nickel, iron, and silicon are added. The nickel increases the strength of the alloy to commercially acceptable levels. Silicon bronze contains as much as 3 percent silicon along with iron, manganese, tin, and zinc. Beryllium bronze with as much as 2.75 percent *beryllium* can be hardened by heat treating.

SELECTING WELDING RODS

While welding rods and electrodes may appear alike to the novice, they are not (Fig. 3-4). Each item is specifically designed to do a particular job and is used with a particular type of welding equipment. An electrode must supply filler

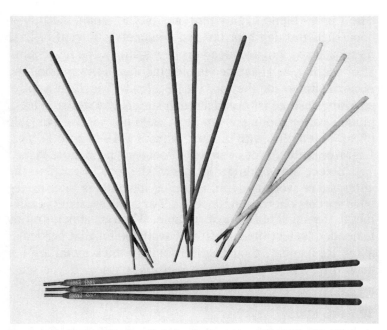

Fig. 3-4. A selection of rods and electrodes (courtesy of Airco Welding Products, Murray Hill, NJ).

metal to the joint being made and a path for electric current. It is the arc at the end, or tip, of the electrode which puddles the working surface and does the work. A welding rod's job is to deposit enough filler metal of the proper composition (alloy) to complete the weld. The rod may also be flux-coated to cut down on oxide deposits, though many rods are also available as bare rods. The flux, if needed, is then brushed on the working surface of the metal being welded.

Generally, thin sheet metal, up to about ⅛ inch, can be welded without the addition of filler metal. The base metal that melts and forms a puddle becomes the filler metal for the joint. From ⅛ inch on up, filler metal of an alloy as close as possible in composition to the base metal is needed. This selection process, along with determining the correct rod size, is one in which you need to gain experience. A correct welding rod will provide, with a neutral flame (oxyacetylene flame characteristics will be covered later), a rippled surface.

The ripples blend evenly into one another and penetrate to about half the depth of the rod's diameter. Fusion of base metal and rod will produce very few sparks. The fused joint, after cooling, will have no visible pinholes. If the welding rod is not suitable for the job, the results will include a lot of sparking during fusion and a rough surface, as well as a lot of pinholes after cooling down. Most welding rods are available in 3-foot lengths, with diameters from 1/16 inch to ¼ inch. Cast iron welding rods come in 2-foot lengths in most cases.

Select the welding rod to suit the job. Determine the alloy you're working with, and then look at the joint to see what sort of penetration is needed. For small amounts of filler metal, the small rod is most suitable. When great penetration is needed, select the largest rod available for that particular alloy. Remember, though, that welds in thick metal are not made in a single pass, but in multiple passes. If you must penetrate ½ inch, a ¼-inch welding rod will do the job.

BRAZING RODS

Brazing or braze welding doesn't require the intense heat of fusion welding, and a weaker joint is produced. Braze welding is often the only answer to the need for strong joints, where application of enough heat to fuse the base metal and welding rod would warp or distort the object being welded. It proves a method of joining dissimilar metals. A variety of rods is used, with the standard brazing rod consisting of a bronze alloy including copper, iron, manganese, tin, silicon, lead, aluminum, chromium, and zinc. These rods can be used to braze weld steel, cast iron, brass, and bronze. When a higher strength filler metal is needed, a manganese bronze rod will usually be your best choice. Manganese bronze rods can be used when braze welding malleable cast iron or steels. For even greater joint strength, a nickel silver brazing rod can be used in almost all applications where a bronze brazing rod will work. Special low temperature brazing rods are available for situations where you have to worry about heat distortion or heat cracking of the work surfaces.

Braze welding of mild steel will provide a tensile strength of about 50,000 psi (Airco's No. 22 bronze rod), while a fusion weld done with Airco's No. 1 alloy steel rod will give you a minimum joint strength of about 60,000 psi. Because of the possibilities of heat distortion and internal metal grain changes, I would recommend braze welding for applications including stainless steels, cast irons, and most copper alloys. Nickel silver rods, such as those from Eutectic, produce an ultimate tensile strength of as much as 85,000 psi, with a melting temperature of only 1050 degrees Fahrenheit. You seldom need to take a chance on ruining the base metal.

FUNCTION OF FLUXES

Many of the operations covered in this book require you to use *fluxes* to prevent oxidation and the inclusion of dirt in the weld. The basic job of any flux is the elimination of as many oxides and as much air as possible from the areas being joined. Flux may come as a coating on the welding or brazing rod or as a separate item, which is then placed on the work surfaces or the welding or brazing rod. Some metals, such as mild steel, have a high enough melting temperature to allow the heat of fusion welding to burn off most dirt and oxides. Many welders don't bother to scrape off rust when fusion welding mild steel, but I feel that's a mistake. Cast iron and other metals don't work out too well that way. With these other metals, a flux is needed to bring oxides contained in the base metal to the surface—cast irons in particular—and to help prevent other oxides from forming. The chemical composition of the flux helps to lower the melting temperature of the oxides so they can be dispersed where they will do no harm to the weld. Dispersion usually is effected by the oxides floating to the surface of the weld. This oxide content forms the slag on the surface of the cooled weld. Slag can be chipped or brushed away once the joint has cooled sufficiently.

Fluxes, like welding rods and electrodes, are chosen to match the job being done. The flux must be correct for the rod being used as well as for the base metals being joined. Oxyacetylene and other gas welding will often require a different flux than that needed for arc welding.

Fluxes are divided into temperature groupings. There are high temperature fluxes for use with copper and brass alloys when you're braze welding at temperatures in excess of 1500 degrees Fahrenheit (816 degrees centigrade). For lower temperatures, a special purpose flux is required. A general purpose type is used when welding at 400 degrees Fahrenheit.

WELDING ELECTRODES

Welding electrodes meet most of the same criteria as welding rods. The electrode provides a path for electric current in order to form the working electric arc. I believe the American Welding Society has done a better job of simplifying the classifications, making the selection of the correct electrode for any job a lot easier. The system of numbers is set up such that you need only know what alloy you are welding to check the number list and get the correct alloy rod.

Flux-coated electrodes and uncoated electrodes are available in most specification ranges. Electrode fluxes are covered at the end of the chapter.

American Welding Society specifications include E60XX electrodes. These are types like the E6010 and E6011 fast freeze electrodes in which the weld solidifies very rapidly, making such electrodes especially good for vertical and overhead uses. The AWS specifications include minimum tensile strength requirements for electrodes of different kinds. The E60XX series must have a tensile strength of 60,000 psi and most, if not annealed or stress relieved by peening, exceed this by as much as 15,000 psi. The E6010 is a dc electrode, while the E6011 is used with an ac arc welder. Both are used with mild steel alloys. The E60XX electrodes are generally more useful in production welding applications than in do-it-yourself jobs.

Low hydrogen electrodes such as the E6015 (dc, reverse polarity) are used on high tensile strength steels and provide greater ductility. E6016 electrodes are used for ac and dc with straight polarity.

Iron powder electrodes provide for a slightly faster method of welding. The coating on the electrode may range from 3 to 60 percent iron powder. The electrodes are best used where a heavy weld deposit needs to be laid down quickly. The more iron powder in the coating, the less adaptable the electrode will be to overhead and vertical work. While still a fast freeze electrode, the fast fill feature will soon develop practical freeze limits. The filler material will tend to sag or drip in such applications. Iron powder provides an electrode that gives a very smooth, attractive bead. The E6014 electrode gives you a shallow penetration in mild steel, with an excellent bead appearance for fillet welds (ac, dc, or either polarity).

For other electrodes such as the E70XX, E80XX, E90XX, and E100XX series, follow the manufacturers' directions. These are alloy or special purpose electrodes, with current needs and most applicable position uses recommended by the makers of the electrodes. As an example, Hobart's LH-818-N1 electrode meets the specifications that the AWS sets up for E8018-C1. This electrode is intended for use in welding nickel alloyed steels where metal toughness is important. The electrode provides a tensile strength of 88,500 psi and is used, in the ⅛ inch size, within a 90 to 160 ampere range. The optimum average is 140 amperes for flat work. For vertical work, the range is 90 to 135 amperes, and for overhead welding the range goes back to 90 to 160 amperes—ac, dc, reverse, or straight polarity. Arc technique differs from the E6010 electrode types since a short arc is needed. A whipping motion cannot be used, though weaving is possible if you use an extremely short arc. Larger electrodes of this type are recommended only for flat and horizontal fillet welding.

Storage becomes something of a problem. Storage temperature should be from 250 to 400 degrees Fahrenheit (121 to 205 degrees centigrade). The electrodes can be reconditioned by heating them to about 600 degrees Fahrenheit (316 degrees centigrade) and holding them there for an hour. These special purpose electrodes are seldom needed. When they are, simply buy as many as you need for the job. Any interim holding or storage can be done in your oven.

SPECIAL PURPOSE NONWELDING ELECTRODES

There is a similarity in structure with cast iron, stainless steel, and other special electrodes that makes it easily possible for you to get excellent results with them just by following the manufacturers' recommendations. Also, follow good welding procedures.

Some electrodes are not intended to produce joints. These electrodes include hard surfacing electrodes, chamfering electrodes, and gouging electrodes. Such special purpose electrodes can prove exceptionally handy around the home and particularly the farm. A series of beads laid down with a hard surfacing electrode on a mower blade or a plow point will give you wear that will far exceed the original wear resistance of the tool. Various types of electrode alloys allow you to build up strong, durable surfaces on everything from thin shovel blades to backhoe bucket teeth and even larger items. Several varieties are manufactured. Some are made to resist severe impact; others resist abrasion and corrosion. For maximum abrasion resistance, you'll need to purchase a chrome carbide tube electrode which is filled with carbide particles. *ChromCarb N*6006* is one brand. For heat resistance, Eutectic Corporation can supply you with *EutecTrode 6800*. The EutecTrode 4 is designed to let you lay down a surface with very high impact resistance. Hobart, Lincoln, Airco and other companies manufacture hard surfacing and buildup electrodes.

Many of these electrodes are used to put down the surface in layers. Hobart's *Tufenhard 550* will give you a Rockwell hardness of 51 on the first pass or layer, with the third pass resulting in a Rockwell hardness of 54 on mild steel. The single layer types, Tufenhard 150 and 160, start out with a much lower hardness and then increase to a Rockwell hardness of 50 as the work hardens. Generally, the higher the Rockwell hardness, the greater the abrasion resistance of the resulting surface. Those hard surfacing alloys intended to increase impact resistance will start out at lower Rockwell ratings before work hardening. Work hardening means that as the tool is used, the surface increases in hardness with the use.

Because arc welders need special accessories for metal cutting, different electrodes have been developed to provide you with specialized tools to do metal preparation jobs that would otherwise require the carbon arc torch. Eutectic Corporation's *Xuper Exotrode* is one. It is used to chamfer and gouge all metals and is specifically designed for use with light duty arc welders. Gouging and chamfering are speedy ways to prepare a crack in an engine block for the welding or braze welding needed to repair it. For piercing stainless steel and cast iron, *CutTrode #1* is good. It can also be used to chamfer, gouge, and cut all metals. Your arc welder, though, must have the dc, straight polarity capability. ChamferTrode is meant for removing old welds and for cutting out rivets, but it also can be used for chamfering and gouging.

ELECTRODE FLUXES

Electrodes require fluxes with different compositions just as welding rods do. Some are dipped during manufacturing, while others are *extruded*—that is, the coating is forced onto the electrode wire as it moves through a press containing a die. Light dippings provide semicoated electrodes, while heavily coated electrodes must be dipped several times or extruded.

Cellulose coatings are used on all-position electrodes. Cellulose coatings are usually used with dc reverse polarity electrodes.

Other compounds are quickly taking the place of cellulose as they provide greater weld cleanliness. Titanium dioxide is one compound tht works well. It is the primary mineral coating for electrodes though other compounds, including sodium silicates and *ferromanganeses,* are also being widely used. In addition to easier welding and cleaner welds, these coatings allow for the use of ac or dc, either polarity.

Iron oxide fluxes are used for a smooth appearance. Weld strength is lowered somewhat.

For low-hydrogen electrodes, calcium carbonate compounds are used as fluxes, sometimes with iron powder added. It is essential that these electrodes be kept very dry.

Special purpose welding jobs will require the use of other elements and compounds, and many are available. If you come up with any difficult or special jobs, a check with your local welding distributor may be able to make your job easier and the result stronger.

Selecting either an electrode or welding rod is a reasonably simple job. You will use mild steel rods or welding electrodes for the majority of your work. Selection of the other types is simplified by the AWS specifications and manufacturers' recommendations, as long as you have an idea of approximately what alloy of what base metal you are working with.

Chapter 4
Gas Welding Equipment

In some ways *gas welding* is a versatile method. The reason for the versatility is simply that gas welding equipment tends to be more portable than arc welding gear. Obviously, arc welding setups are available in portable versions. The need for gas-driven generators to run the equipment adds enormously to price, as does the necessity for a truck or trailer on which to mount the heavy equipment. A complete oxyacetylene setup will cost hundreds of dollars less than a portable arc welder and will usually supply much greater welding and cutting capacity.

The lack of versatility comes when special arc welding equipment, such as tungsten inert gas accessories, is needed to get welds on stainless steels and aluminum. The original cost of a gas welding setup is considerably higher than the cost for arc welding equipment, assuming you already have a home or farm circuitry able to handle the arc welder. I'm talking only about ac arc welders which run about $200 less than a comparable gas welding outfit. Don't forget the need to lease or buy oxygen and acetylene cylinders. Note that ac-dc arc welders, with tungsten inert gas acessories (but without the argon cylinder), will cost about $150 more than a comparable gas welding setup. Prices are taken from the latest Sears, Roebuck and Company tool catalog. All models considered are top of the various lines.

You may wish to add simple, lightweight gas welding equipment to a moderate range arc welder to improve portability, while extending the range of welding jobs possible. You can select oxypropane or oxy-MAPP gas welders of a grade suitable for light welding—considered to be no more than ¼ inch capacity—and add them to your arc welding equipment at expenditures of not much over $200. The oxygen cylinder is included, and MAPP or propane is bought in standard bottles and attached.

Airco's Tote-Weld is a fine outfit. The entire package can't weigh much more than 15 pounds. It does an excellent job of welding, brazing, and cutting. Even lighter tools such as the Solidox torch from Cleanweld and the Bernzomatic oxy-propane torch provide portability at a cost usually under $40.

Some experts predict the near complete demise of gas welding of most styles by the end of the century. I tend to doubt this will happen for oxyacetylene and oxy-MAPP setups offer greater use and adaptability for cutting procedures at reasonable cost. Most oxyacetylene kits available from companies such as Harris, Linde, Airco, Sears, and Montgomery Ward come with a cutting torch head. The cutting capacity, in mild steel, is often 6 inches. Carbon arc and electrode cutting with an arc welder is more limited and can, for extensive use, require additional equipment.

GASES

Fuel gases used with oxygen in gas welding differ in many respects. Acetylene, propane, MAPP gas (a stabilized form of acetylene), butane, and natural gas are commonly used. Methane is also used occasionally. For the purposes of this book, only acetylene, propane, and MAPP gas are covered. These gases are the most important ones for the do-it-yourself welder. Propane is only lightly covered. It is the least popular of the three gases for welding purposes, and the one providing the lowest heat when combined with oxygen.

Acetylene

Acetylene is probably the best known of the welding fuel gases. It produces the hottest flame when combined with oxygen. That flame is about 5,600 degrees Fahrenheit (3095 degrees centigrade). Acetylene is produced by combining hydrogen and carbon. It is very highly explosive in combination with air. Air is not always essential to set acetylene off. If pressures are too high (any thing over 28 pounds per square inch), the gas breaks down into its basic elements and may then make a large bang with virtually any form of ignition.

For welding use, acetylene will *never* be used at pressures greater than 15 pounds per square inch. The gas is available in special cylinders in a variety of sizes. It is unstable and has explosive limits (percent of acetylene) of 3 to 90 percent in oxygen and 2½ to 8 percent in air. Acetylene has a tendency to backfire, but it has low toxicity.

MAPP Gas

MAPP gas is a stabilized version of acetylene. It has a working temperature, when combined with oxygen, of about 5,300 degrees Fahrenheit (2929 degrees centigrade). It is stable at all normal pressures and doesn't need the internal stabilizers inside the cylinder that acetylene must have (usually either acetone and porous cement or balsa wood). MAPP gas has a small tendency to backfire. It has an allowable regulator pressure of 225 psi at temperatures as high as 130 degrees Fahrenheit (54.5 degrees centigrade). The Btu (British thermal unit) per pound output is comparable to that of acetylene. MAPP gas produces 21,100 Btus per pound, while acetylene supplies 21,500 Btus. MAPP gas is explosive in concentrations of 3.4 to 10.8 percent in air and 2.5 to 60 percent in oxygen.

Propane

Propane is useful in welding and cutting outfits, though it simply doesn't provide the same temperature given by MAPP gas and acetylene. It is actually much more useful in single fuel torches used for soldering jobs. Propane in combination with oxygen will produce a temperature of about 4550 degrees Fahrenheit (2512 degrees centigrade). It also requires about five times as much oxygen to maintain the same volume of heat as is produced by MAPP gas or acetylene. The Btu per pound output is quite high at 21,800, but not very impressive when you consider its oxygen requirements and low flame temperature. Propane is classified as stable when subjected to shock. It has explosive limits of 2.4 to 57 percent in oxygen and 2.3 to 95 percent in air. Backfire tendency is low, as is toxicity. Propane causes few adverse reactions in common materials (Table 4-1).

Oxygen

Commercially pure oxygen, as used in welding, is guaranteed to be of a minimum 99.5 percent pure, with about .5 percent nitrogen. It is a nonflammable gas that enters into combinations with flammable portions of other compounds (not just gases) and supports the resulting combustion. Because there are many compounds that often combine with oxygen, causing rapid combustion, great care must be used to isolate the oxygen. Pressures up to 2,200 pounds per square inch are almost standard for oxygen storage, so care is needed with the storage vehicles. Gases used with arc welding procedures (MIG and TIG) are covered in their appropriate chapters.

These gases are stored at high pressures. While a cracked regulator neck might not immediately cause a fire, the damaged cylinder can easily take off like a rocket, causing major damage and injury over a wide area. The pressure at even a minor crack could cause a cylinder to explode, even without the flame ignition needed by stable gases.

CYLINDERS

Because of the pressure storage needs of fuel gases, the cylinders must be treated carefully and stored in a secure, upright position. Lower pressure acetylene cylinders are to be stored upright and protected against shock.

An acetylene cylinder is filled with a mixture of porous concrete. Then acetone is added, in which the acetylene dissolves. Not all internal acetylene fillings are concrete. Often long fiber asbestos is used at the top, with fine asbestos at the bottom with a central filler of balsa wood. At the bottom

Table 4-1. Comparison of Gases Used in Welding (courtesy of MAPP Products, Airco Welding Products, Murray Hill, NJ).

	MAPP gas	acetylene	natural gas	propane	propylene
SAFETY					
Shock sensitivity	Stable	Unstable	Stable	Stable	Stable
Explosive limits in oxygen, %	2.5-60	3.0-93	5.0-59	2.4-57	2.3-55
Explosive limits in air, %	3.4-10.8	2.5-80	5.3-14	2.3-9.5	2.0-11.1
Maximum allowable regulator pressure, psi (kPa)	Cylinder	15 (103)	Line	Cylinder	Cylinder
Burning velocity in oxygen, ft/sec (mm/sec)	15.4 (4694)	22.7 (6097)	15.2 (4633)	12.2 (3718)	15.0
Tendency to backfire	Slight	Considerable	Slight	Slight	Slight
Toxicity	Low	Low	Low	Low	Low
Reactions with common materials	Avoid alloys with more than 67% Cu	Avoid alloys with more than 67% Cu	Few restrictions	Few restrictions	Few restrictions
PHYSICAL PROPERTIES					
Specific gravity of liquid, (60/60 F)	0.576	—	—	0.507	0.5220
Lb/gal liquid at 60 F (kg/, m³ at 15.6 C)	4.80 (575)	—	—	4.28 (513)	4.35 (526)
Ft³/lb gas at 60 F (m³/kg at 15.6 C)	8.85 (0.55)	14.6 (0.91)	23.6 (1.4)	8.66 (0.54)	8.25 (0.51)
Specific gravity gas (air=1) at 60 F (16.6 C)	1.48	0.906	0.62	1.52	1.476
Vapor pressure at 70 F psig (20 C, kPa)	94 (648)	—	—	120 (827)	135 (1561 at 37.8C)
Boiling range. F (C) 760 torr	-36 to -4 (-36 to -20)	-84 (-64)	-161 (-107)	-50 (-45.6)	-53.86 BP (-47.7)
Flame temperature in O₂. F (C)	5301 (2927)	5589 (3087)	4600 (2538)	4679 (2526)	5193 (2867)
Latent heat of vaporization at 25 C, Btu/lb (kJ/kg)	227 (528)	—	—	184 (428)	188 (437)
Total heating value (after vaporization) Btu/lb (kJ/kg)	21,000 (49,000)	21,500 (50,000)	23,900 (56,000)	21,800 (51,000)	21,100 (49,000)
Btu/ft³ (MJ/M³)	2406 (90)	1470 (55)	900-1000 (34-37)	2498 (93)	2371 (88)

of all acetylene cylinders you'll find a steel band to protect the base of the cylinder, and there are two or more safety plugs in that base. At the top of the cylinder, as a part of the valve base, is another safety plug. These safety plugs are filled with a low melting temperature alloy so that, in case of fire, the acetylene will escape through the plugs instead of blowing the cylinder apart. Keep your torch flame well away from the cylinder at all times. The standard procedure is to fill acetylene cylinders to 250 psi, with a resulting line pressure (working pressure) kept to 15 pounds by the regulator. There is no way to keep acetylene dissolved in acetone once it leaves the cylinder.

MAPP gas cylinders do not need the elaborate precautions built into acetylene cylinders. Thus, they are most usually similar to many steel cylinders built to withstand similar pressures. Propane is stored in just about the same kind of cylinder. Again, these cylinders should be protected against flame and shock.

Oxygen cylinders are made of seamless drawn steel as are the previous two kinds of cylinders. Normally an oxygen cylinder will be painted red or green—most often green—as a distinguishing mark. Oxygen is stored at very high pressures, usually about 2200 psi for welding purposes. Care is needed while handling the cylinders. The most common size is probably the 244-cubic-foot cylinder which weighs about 150 pounds when full (20 pounds less when empty). Half-size cylinders are available.

Most oxygen and fuel gas cylinders are rentals. Long-term rentals are best.

Oxygen cylinders must be kept away from all types of greases and oils made from petroleum, as combining them may result in a bad explosion. Again, all cylinders must be handled carefully. The torch's flame, hot slag, and sparks should be kept away from them.

REGULATORS

Acetylene and MAPP gas are stored at high pressures. *Regulators* of various types are used to reduce the pressure of both oxygen and fuel gases to working levels. They help prevent fluctuation of gas pressure at the torch tip. Welding gas regulators must be sensitive enough to keep the pressure, at the torch tip, to within just a few ounces per square inch of optimum. They must also be strong enough to withstand pressure reduction from 2200 psi to as little as 10 psi or less. The regulators must be accurate enough to maintain the correct level of pressure at the torch even though the cylinder pressure, especially over several hours of welding, may drop as much as 50 percent. The regulators must be adjustable over a fair range. One job may require only 10 psi of oxygen, while another might require as much as 20 psi. They must be rugged enough to withstand constant use and hauling around the workshop (Fig. 4-1).

A *single-stage* regulator will generally have a spring-controlled diaphragm. The spring is adjustable by some sort of screw device to give the appropriate pressure as indicated on the gauge. The gas flows into a chamber and is reduced in pressure as the diaphragm closes off all gas over 10 psi. When the pressure starts to drop below 10 psi, or whatever pressure has been dialed into the regulator, the diaphragm will reopen and allow more gas to flow until the proper pressure is maintained. The screw adjustment allows an almost infinite pressure adjustment from zero to full cylinder pressure (Fig. 4-2).

Two-stage regulators are more accurate, but they are not totally essential to normal welding operations where precise metering and control are not of great importance. Bill Dreifke at Harris Calorific believes that the single-stage regulator is sufficient for virtually all home, farm, and hobby welding jobs, as well as most commercial and construction applications. The basic difference in single-stage and two-stage regulators is the greater pressure stability at the torch tip. The first regulator chamber in a two-stage regulator drops pressure to about 220 psi (on a fully charged cylinder) and lets the second chamber make a more precise adjustment of the final pressure before the gas flows to the torch tip. The first reducing chamber will usually be fitted with a preset diaphragm spring combination that reduces pressure a set amount. The second reducing chamber is then used as the adjustable, so that the

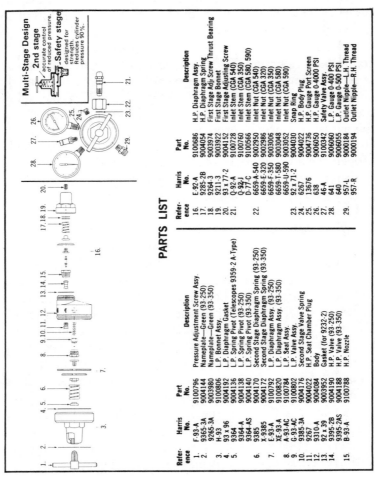

Fig. 4-1. Parts of the 93 multi-stage cylinder regulator (courtesy of Harris Calorific Division).

Multi-Stage Design
2nd stage — accurate control of reduced pressure.
Safety stage — designed for strength. Reduces cylinder pressure 90%.

PARTS LIST

Reference	Harris No.	Part No.	Description
1.	F-93-A	9100796	Pressure Adjustment Screw Assy.
2.	9365-3A	9004144	Nameplate—Green (93-250)
	9265-3A	9003980	Nameplate—Green (93-350)
3.	H-93	9100806	L.P. Bonnet Assy.
4.	93 x 96	9004192	L.P. Diaphragm Gasket
5.	9364	9004136	L.P. Spring Pivot (Telescopes 9359-2 A-Type)
	9364-A	9004138	L.P. Spring Pivot (93-250)
	9364-AS	9004140	L.P. Spring Pivot (93-350)
6.	9385	9004170	Second Stage Diaphragm Spring (93-250)
	X-9385	9004172	Second Stage Diaphragm Spring (93-350)
7.	E-93-A	9100792	L.P. Diaphragm Assy. (93-250)
	XE-93-A	9100820	L.P. Diaphragm Assy. (93-350)
8.	A-93-AC	9100784	L.P. Seat Assy.
9.	G-93-AC	9100802	L.P. Valve Assy.
10.	9385-3A	9004176	Second Stage Valve Spring
11.	9267	9004022	H.P. Seat Chamber Plug
12.	9310-A	9004084	Body
13.	92 x 39	9003952	Gasket (for 9232-2)
14.	9395-2B	9004190	H.P. Valve (93-250)
	9395-2AS	9004188	H.P. Valve (93-350)
15.	B-93-A	9100788	H.P. Nozzle
16.	E-92-A	9100686	H.P. Diaphragm Assy.
17.	9285-2B	9004054	H.P. Diaphragm Spring
18.	9264-3	9003974	First Stage Adj. Screw Thrust Bearing
19.	9211-3	9003922	First Stage Bonnet
20.	93 x 77-2	9004152	First Stage Adjusting Screw
21.	Q-92-A	9100728	Inlet Stem (CGA 540)
	9-92-J	9100760	Inlet Stem (CGA 350)
	D-77-C	9100566	Inlet Stem (CGA 580, 590)
22.	6659-A-540	9002950	Inlet Nut (CGA 540)
	6659-E-320	9002986	Inlet Nut (CGA 320)
	6659-F-350	9003006	Inlet Nut (CGA 350)
	6659-T-580	9003048	Inlet Nut (CGA 580)
	6659-U-590	9003052	Inlet Nut (CGA 590)
23.	92 x 71-2	9004030	Snap Ring
24.	6267	9004022	H.P. Body Plug
25.	13676	9004736	H.P. Gauge Port Screen
26.	638	9006050	H.P. Gauge 0-4000 PSI
27.	46-A	9100342	Safety Valve Assy.
28.	641	9006060	L.P. Gauge 0-400 PSI
	640	9006055	L.P. Gauge 0-500 PSI
29.	957-L	9000184	Outlet Nipple—L.H. Thread
	957-R	9000194	Outlet Nipple—R.H. Thread

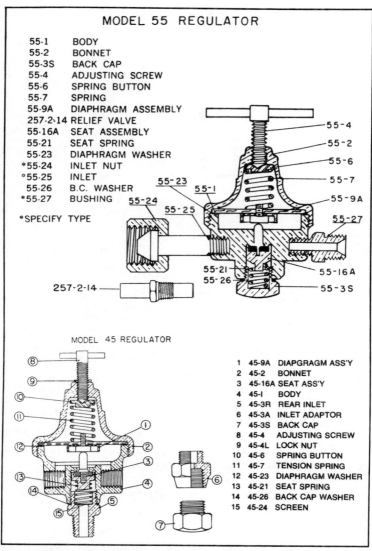

MODEL 55 REGULATOR

55-1	BODY
55-2	BONNET
55-3S	BACK CAP
55-4	ADJUSTING SCREW
55-6	SPRING BUTTON
55-7	SPRING
55-9A	DIAPHRAGM ASSEMBLY
257-2-14	RELIEF VALVE
55-16A	SEAT ASSEMBLY
55-21	SEAT SPRING
55-23	DIAPHRAGM WASHER
*55-24	INLET NUT
*55-25	INLET
55-26	B.C. WASHER
*55-27	BUSHING

*SPECIFY TYPE

MODEL 45 REGULATOR

1	45-9A	DIAPGRAGM ASS'Y
2	45-2	BONNET
3	45-16A	SEAT ASS'Y
4	45-I	BODY
5	45-3R	REAR INLET
6	45-3A	INLET ADAPTOR
7	45-3S	BACK CAP
8	45-4	ADJUSTING SCREW
9	45-4L	LOCK NUT
10	45-6	SPRING BUTTON
11	45-7	TENSION SPRING
12	45-23	DIAPHRAGM WASHER
13	45-21	SEAT SPRING
14	45-26	BACK CAP WASHER
15	45-24	SCREEN

Fig. 4-2. Single-stage regulator (courtesy of Dockson Corporation).

lower pressure and lesser reduction needed may allow for more precise pressure control at the torch (Fig. 4-3). Because of the two-stage effect, these regulators seldom suffer from icing of their working parts, which is an occasional problem with the single reduction stage of single-stage regulators.

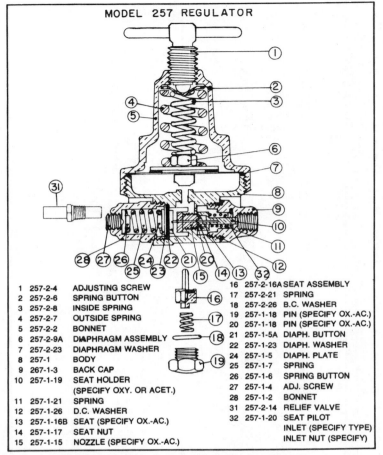

MODEL 257 REGULATOR

1	257-2-4	ADJUSTING SCREW
2	257-2-6	SPRING BUTTON
3	257-2-8	INSIDE SPRING
4	257-2-7	OUTSIDE SPRING
5	257-2-2	BONNET
6	257-2-9A	DIAPHRAGM ASSEMBLY
7	257-2-23	DIAPHRAGM WASHER
8	257-1	BODY
9	267-1-3	BACK CAP
10	257-1-19	SEAT HOLDER (SPECIFY OXY. OR ACET.)
11	257-1-21	SPRING
12	257-1-26	D.C. WASHER
13	257-1-16B	SEAT (SPECIFY OX.-AC.)
14	257-1-17	SEAT NUT
15	257-1-15	NOZZLE (SPECIFY OX.-AC.)

16	257-2-16A	SEAT ASSEMBLY
17	257-2-21	SPRING
18	257-2-26	B.C. WASHER
19	257-1-18	PIN (SPECIFY OX.-AC.)
20	257-1-18	PIN (SPECIFY OX.-AC.)
21	257-1-5A	DIAPH. BUTTON
22	257-1-23	DIAPH. WASHER
24	257-1-5	DIAPH. PLATE
25	257-1-7	SPRING
26	257-1-6	SPRING BUTTON
27	257-1-4	ADJ. SCREW
28	257-1-2	BONNET
31	257-2-14	RELIEF VALVE
32	257-1-20	SEAT PILOT
		INLET (SPECIFY TYPE)
		INLET NUT (SPECIFY)

Fig. 4-3. Two-stage regulator (courtesy of Dockson Corporation).

Welding regulators are generally designed to deliver gas pressures from 50 to 75 pounds, with a maximum delivery pressure as high as 200 psi. Specialized cutting gauges for commercial use may meet demands of more than twice that 200 psi level. The oxygen gauge is usually calibrated to 100 psi, though heavy-duty models such as Harris Calorific's model No. 25 are calibrated to 150 psi. Acetylene regulators will have cylinder pressure gauges reading up to 400 psi, with working pressure gauges reading to 30 psi in the heavy-duty models (Fig. 4-4).

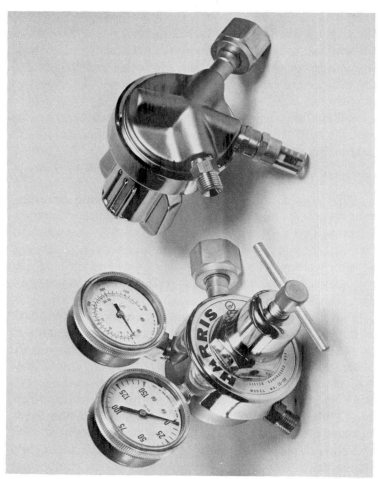

Fig. 4-4. The Harris Model 25 regulator for oxygen, with working pressure calibrated to 150 psi (courtesy of Harris Calorific Division).

Oil is never used to lubricate any of the fittings on the equipment which might come in contact with oxygen. All welding equipment is built of materials that seldom need lubrication. Specialized lubricants which do not contain any petroleum products are used.

HOSES

Gas welding *hoses* are specifically made for the gases used in welding and cutting. One hose, usually green, will be used to transport oxygen, while a red hose will transport acetylene or whatever fuel gas you are using. Hose size should be matched to tip and torch size—the larger the torch, the larger the hose. Follow manufacturer's recommendations for hose size selection (flow recommendations from the hose manufacturers will indicate both torch size and hose length for appropriate hose brands). Hose connections consist of nipples and nuts, with the oxygen nut having a right-hand thread and the fuel gas nut a left-hand thread. Acetylene or fuel gas nuts also have a groove cut around their centers to help prevent temporary mix-ups (Fig. 4-5).

WELDING TORCHES

Each *welding torch* will have a handle with two inlet connections at the base end where the gases are to enter. There will be a valve at each of these inlets so that gas flow can easily be controlled. Two primary types of gas torches are used for welding—the *injector* style for low pressure use and the *balanced,* or equal, *pressure* style for medium pressure use. There are three reasons why the balanced or equal pressure type is better. First, it is not as subject to flashback as the injector style. Second, it is more easily adjustable to provide the most suitable flame for a job. Third, it is readily available.

The torch you see in Fig. 4-6, a *Dockson model 5,* is a balanced pressure type. It is shown with its cutting attachment and a series of tips to provide different size flames. Each balanced pressure type of torch has a handle, a rear section with two tubes for the gases running into the torch, two needle valves, a mixing chamber, and a welding tip. Actual internal construction varies. Some companies produce welding torches that have the oxygen and fuel gas tubes running side by side through the handle. Others make torches with the oxygen tube carried inside the fuel gas tube. Once the gases

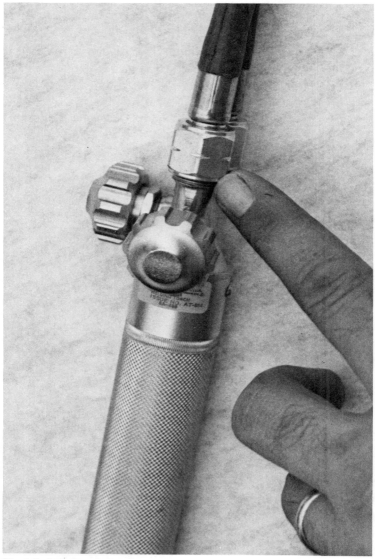

Fig. 4-5. Acetylene connector.

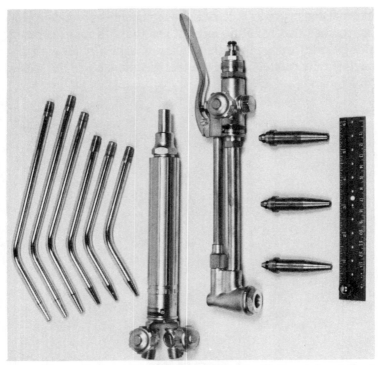

Fig. 4-6. The Dockson model 5 torch and cutting attachment (courtesy of Dockson Corporation).

reach the mixing chamber, they are combined and fed to the torch tip where they are lighted to produce your welding flame.

WELDING TIPS

Welding *tips* come in many sizes. The size of the tip is determined by the size of the hole drilled in it to allow passage of the mixed gases, but this is not always a good way to determine the actual tip performance (Fig. 4-7). The reason is simply because differing internal designs of the torches can create different gas velocities, thus resulting in different performance with the same size tip orifice if used on different outfits. Welding tips vary in orifice size from a 000, suitable for metals to about 3/32 inch thick, on to a 12 for metals 4 inches thick. Actual tip size for any job is determined

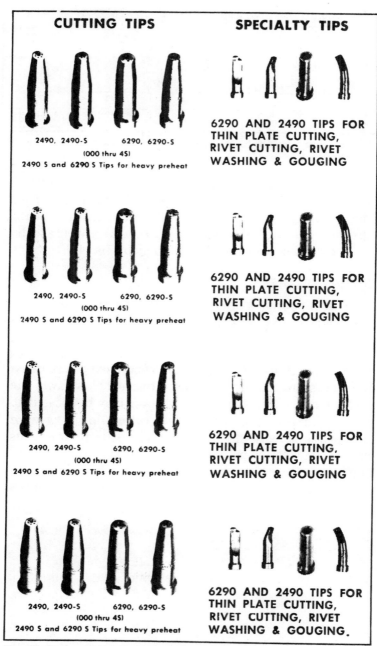

CUTTING TIPS

SPECIALTY TIPS

2490, 2490-S   6290, 6290-S
(000 thru 4S)
2490 S and 6290 S Tips for heavy preheat

6290 AND 2490 TIPS FOR THIN PLATE CUTTING, RIVET CUTTING, RIVET WASHING & GOUGING

2490, 2490-S   6290, 6290-S
(000 thru 4S)
2490 S and 6290 S Tips for heavy preheat

6290 AND 2490 TIPS FOR THIN PLATE CUTTING, RIVET CUTTING, RIVET WASHING & GOUGING

2490, 2490-S   6290, 6290-S
(000 thru 4S)
2490 S and 6290 S Tips for heavy preheat

6290 AND 2490 TIPS FOR THIN PLATE CUTTING, RIVET CUTTING, RIVET WASHING & GOUGING

2490, 2490-S   6290, 6290-S
(000 thru 4S)
2490 S and 6290 S Tips for heavy preheat

6290 AND 2490 TIPS FOR THIN PLATE CUTTING, RIVET CUTTING, RIVET WASHING & GOUGING.

Fig. 4-7. Cutting tips for Harris torches (courtesy of Harris Calorific).

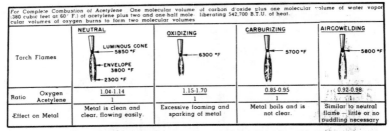

For Complete Combustion of Acetylene One molecular volume of carbon dioxide plus one molecular volume of water vapor .380 cubic feet at 60° F.) of acetylene plus two and one-half molecular volumes of oxygen burns to form two molecular volumes liberating 542,700 B.T.U. of heat.			
	NEUTRAL	OXIDIZING	CARBURIZING	AIRCOWELDING
Torch Flames	LUMINOUS CONE 5850 °F ENVELOPE 3800 °F 2300 °F	6300 °F	5700 °F	5800 °F
Ratio Oxygen Acetylene	1.04-1.14 1	1.15-1.70 1	0.85-0.95 1	0.92-0.98 1
Effect on Metal	Metal is clean and clear, flowing easily.	Excessive foaming and sparking of metal	Metal boils and is not clear.	Similar to neutral flame — little or no puddling necessary

Fig. 4-8. Types of welding flames (courtesy of Airco Welding Products, Murray Hill, NJ).

by the position of the weld, the type of metal being welded, the thickness of the metal, and the kind of joint being welded. Much depends on the welder's skill. Other factors affecting tip orifice size are the speed needed to make the weld and the type of welding flame needed for the job (Fig. 4-8).

CUTTING TORCHES

The *cutting torch* is an addition that fits on the base of the oxyacetylene, oxy-MAPP, or oxypropane torch. It provides you with a third tube to supply pure oxygen to the area being cut as soon as the oxyfuel gas has heated it to or near melting temperature (Fig. 4-9). The stream of pure oxygen released under pressure actually does the cutting and blows away the oxides (slag). The oxygen control lever may be located almost anywhere on your torch, but the cutting tip will have the oxyfuel gas holes located, in a circular pattern, around its outside rim with the pure oxygen hole located in the center. Torch styles are the same as are welding torch styles— injector and balanced, or equal, pressure. Cutting head angles

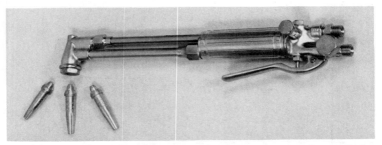

Fig. 4-9. Dockson model CM cutting torch (courtesy of Dockson Corporation).

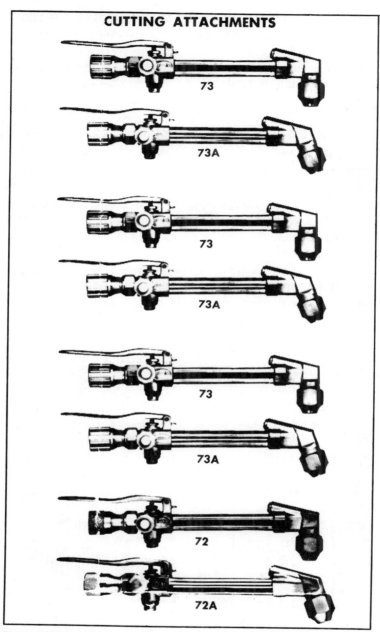

CUTTING ATTACHMENTS

73

73A

73

73A

73

73A

72

72A

Fig. 4-10. Cutting attachments for welding torches (courtesy of Harris Calorific Division).

67

will differ to some degree. Select a head angle that is comfortable for you to use (Fig. 4-10). The cutting head will vary slightly depending on the kind of fuel gas you expect to use, so match the cutting head to the fuel gas you plan to use most often. Head angles available are straight, 75 degrees, and 90 degrees.

CUTTING TIPS

Cutting tips for use with acetylene differ from those used with propane. The propane model has an outer shield which is essential to add stability to the flame. Cutting tips range from a 50VVCM, for cutting plate up to 1/6 inch thick, to an 8HNM for metals as much as 15 inches thick (these are Harris Calorific tip size designations and will not necessarily apply to those of Airco, Dockson, and others (Tables 4-2 and 4-3). Often you'll find torch size and tip orifice size in a packaged gas welding outfit limit cutting to about 6 inches of mild steel, which is likely more than most of you will need to slice through.

In addition to cutting tips, you'll also find heating tips available for gas welding outfits. These are usually only needed if you plan doing much work on metals needing extensive preheating, heat treating, or annealing. Tip size varies according to the requirements of the job.

Table 4-2. Type SP (One Piece) and Type FS (Two Piece) Standard Speed Cutting Tips (courtesy of Airco Welding Products, Murray Hill, NJ).

| plate thk. inches | tip size no. | cutting speed in./min. | Oxygen | | | | MAPP Gas | | kerf width inches |
| | | | Cutting | | Preheat | | | | |
			press P.S.I.G.	flow C.F.H.	press P.S.I.G.	flow C.F.H.	press P.S.I.G.	flow C.F.H.	
3/16	72	24-30	30-40	20-30	5-10	5-20	2-6	2-8	.03
1/4	68	22-28	35-45	30-40	5-10	5-20	2-6	2-8	.035
3/8	65	21-27	35-45	40-55	5-10	5-25	2-8	2-10	.04
1/2	60	20-26	40-50	55-65	5-10	10-25	2-8	4-10	.045
3/4	56	16-21	40-50	60-75	5-10	10-25	2-10	4-10	.06
1	56	14-19	40-50	60-75	5-10	15-25	2-10	6-10	.06
1-1/4	54	13-18	40-60	105-120	10-20	20-38	2-10	8-15	.08
1-1/2	54	12-16	40-60	105-120	10-20	20-38	2-10	8-15	.08
2	52	10-14	40-60	145-190	10-20	20-38	2-10	8-15	.09
2-1/2	52	9-13	40-60	150-200	10-30	20-50	6-10	8-20	.09
3	49	8-11	40-70	200-250	10-30	20-50	6-10	8-20	.10
4	44	7-10	40-70	300-360	10-30	20-50	6-10	10-20	.14
6	44	5-8	40-70	300-360	10-30	20-50	6-15	10-20	.14
8	38	4-6	40-80	415-515	30-50	50-100	10-15	20-40	.17
10	31	3-5	40-90	550-750	30-50	50-100	10-15	20-40	.19
12	28	3-5	40-100	750-950	30-50	75-150	10-15	30-60	.22
14	19	2-4	40-120	1000-1250	30-50	75-150	10-15	30-60	.25

| plate thk. inches | tip size no. | cutting speed in./min. | Oxygen | | | | MAPP Gas | | kerf width inches |
| | | | Cutting | | Preheat | | | | |
			press P.S.I.G.	flow C.F.H.	press P.S.I.G.	flow C.F.H.	press P.S.I.G.	flow C.F.H.	
1/4	68	24-31	60-70	55-65	5-10	8-25	2-10	3-10	.05
3/8	65	23-30	70-80	60-80	5-10	10-25	2-10	4-10	.055
1/2	60	22-29	80-90	75-95	5-10	12-25	2-10	5-10	.06
3/4	56	20-26	80-90	115-130	5-10	12-25	2-10	5-10	.065
1	56	18-24	80-90	115-130	5-10	12-25	2-10	5-10	.065
1-1/4	54	16-22	70-80	155-170	10-20	20-38	2-10	8-15	.08
1-1/2	54	15-20	80-90	170-180	10-20	20-38	2-10	8-15	.08
2	52	14-19	80-90	215-255	10-20	20-38	2-10	8-15	.09
2-1/2	52	12-17	80-90	215-255	10-20	20-38	4-10	8-15	.09
3	49	10-14	80-90	310-365	10-20	20-38	6-10	8-15	.10
4	44	9-13	80-90	420-510	10-20	30-38	6-10	10-15	.12
6	44	7-11	80-90	420-510	10-20	30-38	10-15	10-15	.12
8	38	6-9	80-90	590-720	15-30	30-50	10-15	15-20	.15

CHECK VALVES

Most of the companies making gas welding equipment also produce small safety units called *check valves* (Fig. 4-11). These units are quick closing valves designed to prevent the reverse flow of gases back into the hoses or regulators, where mixing of the fuel gas and oxygen could cause an explosion. I definitely recommend the use of check valves on every gas welding setup, with the valves installed either at the regulator/hose connection or at the torch/hose connection. Protection is greatest with the check valves in place at the torch/hose connection. The check valves cannot be used with either the Solidox or Bernzomatic lightweight welding kits. These check valves are not designed to stop a flame running back down the hose, but according to Harris Calorific the industry strongly believes that about 90 percent of gas welding equipment explosions are caused by the reverse flow of the gases when the heat of recompression drives temperatures up beyond the ignition point.

OTHER GAS WELDING OUTFITS

Several companies, most notably Cleanweld with its Solidox units and Bernzomatic with its OX5000 welders, provide specialized, lightweight gas welding outfits for the do-it-yourself welder. These tools can cut mild steel up to about ¼ inch thick. Both units can use either propane or MAPP gas.

The Solidox torch uses a candle to produce oxygen. The candle is lighted and slipped into a tube in the unit. The tube is then sealed, and within a minute or so there is enough oxygen being produced to give a good, hot flame. There is no way for you to adjust oxygen flow, though, and welding time is limited.

The OX5000 from Bernzomatic uses oxygen at 400 psi in small cylinders. The oxygen cylinders are painted green and have a right-hand thread on the top. Both oxygen and fuel gas rates are adjustable with the OX5000.

The OX5000 is a one-piece outfit. All parts are held in one hand, including the oxygen and fuel gas tanks, while you're doing the job. The Solidox has hoses about 4 feet long. The lighter weight of the hose-fed torch gives somewhat better results on longer welds, cuts, and braze joints. The slight inconvenience of lighting a candle and inserting it in the holder with the Solidox welder counterbalances the ease of use. The longest burning Solidox oxygen candle will provide you with about 12 minutes of oxygen. Several types of candles are available, including an oxygen-rich one to make cutting or mildly heavier welding easier. Some smaller oxygen candles are used with smaller tips for very fine work. The OX5000 has 1.1 cubic feet of oxygen in each cylinder. In most welding operations, you can expect this to last about 15 minutes, but actual operating time is controlled by oxygen flow to the torch tip. For light brazing work, you'll get more time, while for cutting you need to change oxygen cylinders more often. The OX5000 has only one tip available.

Airco makes an oxy-Mapp outfit with a convertible torch. The Tote-Weld is delivered with tips that cut up to ⅜-inch

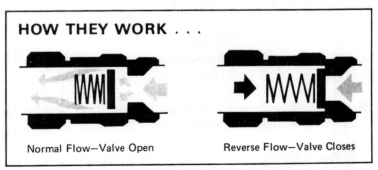

HOW THEY WORK . . .

Normal Flow—Valve Open   Reverse Flow—Valve Closes

Fig. 4-11. How a check valve works (courtesy of Harris Calorific Division).

Fig. 4-12. Tote-Weld welding torch in action (courtesy of Airco Welding Products, Murray Hill, NJ).

steel and weld up to 3/32-inch steel. Tips are available that will more than double the Tote-Weld's capacity for welding, while a 1+ tip will raise cutting thicknesses to about ½ inch. Tote-Weld models come with 20-pound refillable oxygen

Fig. 4-13. Using the Tote-Weld to flame cut (courtesy of Airco Welding Products, Murray Hill, NJ).

cylinders and 1-pound disposable MAPP or acetylene cylinders. The MAPP gas cylinders are much safer than acetylene ones. The Tote-Weld and the Cut 'n' Weld from Sears provide good versatility and portability (Figs. 4-12 and 4-13). A lighter kit will be an ideal supplement to an arc welder.

EQUIPMENT SELECTION

Certain factors make gas welding equipment better for some jobs, while other factors make arc welding equipment handier. Basically, if portability is a major concern, gas welding equipment will usually be cheaper than generator-driven arc welders. Brazing and cutting are more easily carried out with oxyfuel gas equipment.

For stationary use, the arc welder is more functional. The arc is hotter than the gas flame, so you will be able to make a stronger weld in many metals. The arc welder has greater adaptability in overhead, horizontal, and vertical welding applications because an ac-dc welder may allow you to choose between two polarities and current types, with each polarity having different welding characteristics. Arc welders can be adapted, at a fairly reasonable cost, for use with tungsten inert gas welding attachments when welding metals such as aluminum. Cleaner joints in stainless steel can be produced.

A gas torch, though, can preheat cast iron before welding. If your present needs include only light cutting and welding, and much of the work is done away from your workshop, your best bet would be a setup like the Tote-Weld. If you plan to work on older farm equipment, a better bet would be a full size kit such as Harris Calorific's 6325 package or Airco's Super Aircomaster outfit. Either of these packages will give you the ability to weld, with appropriate tips, even fairly heavy mild steels (up to 1 inch with the Harris Calorific kit and about ¼ inch with the tips supplied with the Airco outfit) with much larger cut capacity (Harris Calorific at 6 inches, Airco at 2 inches). Welding and cutting capacities are limited by the tips supplied with the kits, and other tips are available. The Harris Calorific 6325 is a single-stage outfit while the Airco is a two-stage setup. Each includes all the hoses needed plus goggles, spark lighter, and attachment wrench. Cylinders, gases, and a set of check valves are needed. Cost is more than that of a similar capacity arc welding setup because of the cylinders' price whether you lease them or buy them outright.

WELDING ACCESSORIES

A *spark lighter* is an essential addition to any gas welding oufit that doesn't have one. The cost is minimal. No gas welding, cutting, or preheating torch should ever be lighted with a match. You may burn your hand. A spark lighter is easy to use (Fig. 4-14).

Other tools will be needed as your welding progresses to more complex joints. Simple joints may need a temporary support before the first tack weld is in place. Since the pieces being welded often need to be held together as the weld solidifies, a series of different sized clamps for different

Fig. 4-14. Using a spark lighter.

Fig. 4-15. Safe dress for welding is essential (courtesy of Airco Welding Products, Murray Hill, NJ).

configurations will be needed. On some jobs a few firebricks or pieces of scrap steel can be used to make the needed supports for the work.

The *chipping hammer/wire brush* combination tool used to remove slag from welds is a fairly economical device. The combination tools can often be bought with replaceable wire brush inserts like Airco's Model A.

Always wear your goggles and welding gloves. For overhead work, you may need arm protectors, though usually a long-sleeved cotton or wool shirt will provide enough protection (Fig. 4-15).

Some materials for testing the weld are required. Most of these items are moderate in cost, and only small amounts will be needed. Dye penetrant is sprayed on the weld and aids in locating surface cracks and porosity, seams, and other weld problems. It comes in a spray can, and the cost is modest.

A Zyglo fluorescent penetrant kit is used to look for pores, cracks, leaks, and other defects. Forget the magnetic particle detection kits. They are too expensive.

Chapter 5
Gas Welding

Equipment setup is the first step in learning to gas weld. Your complete outfit will consist of a cylinder of oxygen, a cylinder of fuel gas, hoses, regulators, torch, goggles, gloves, spark lighter, and welding rod.

First, attach the oxygen regulator to the oxygen cylinder and the acetylene regulator to its cylinder. Attachments are made only after you check that no oil, grease, or dirt contaminates the threads of the regulators, and tanks. The hoses are now connected to their respective fittings—green to oxygen, red to acetylene or other fuel gas. Connections must be made securely, but without the use of excessive force. The couplings are brass and will strip, but the seating must be firm enough so there are no leaks. Check the fittings for the hose torch end and the torch fittings for contaminants. Attach the torch to the hose, firmly but gently.

When you are setting up the equipment, your hands and clothes must be free of grease and oil. Once the hookups are complete, make sure the cylinders are either attached to the wall (or other suitable solid support) or carried on a suitable cylinder truck.

DETAILED SETUP

Secure both cylinders (Fig. 5-1). Remove the valve protector cap from the oxygen cylinder, and very slightly crack

Fig. 5-1. Properly secured gas and oxygen cylinders are important for safety (courtesy of Airco Welding Products, Murray Hill, NJ).

the valve on the cylinder (Fig. 5-2). Cracking the valve will clearaway any dust or other grime that may have marred the seat of the regulator inlet nipple or been carried into the regulator itself where it could cause problems. Connect the oxygen regulator to the cylinder; remember that you are working with a right-hand thread screw. Use either the wrench supplied with your outfit or a top quality adjustable wrench (never a pipe wrench) to draw the nut up tight. Make certain you don't use so much force that the nut is stripped (Fig. 5-3). If there is a leak after a reasonable amount of force is used, remove the regulator. Clean the nipple seat on the regulator and the nipple seat on the cylinder. Since oxygen may not show its presence by hissing out of the leak, check the connections with a mixture of soapy water (use soap, not detergent) each time the hookup is made (Fig. 5-4). Back out the pressure regulator screw on the regulator, making sure it is loose (Fig. 5-5).

Acetylene and other fuel gas cylinders have left-hand threads so that confusion with oxygen outlets is nearly impossible. Use a T-handle wrench to crack the acetylene valve and reclose it (Fig. 5-6). Connect the acetylene regulator to

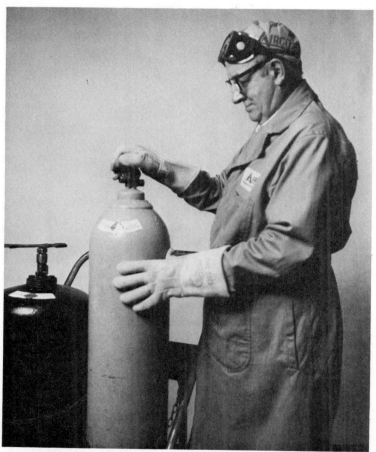

Fig. 5-2. Start by cracking the cylinder valve to clear away any dust (courtesy of Airco Welding Products, Murray Hill, NJ).

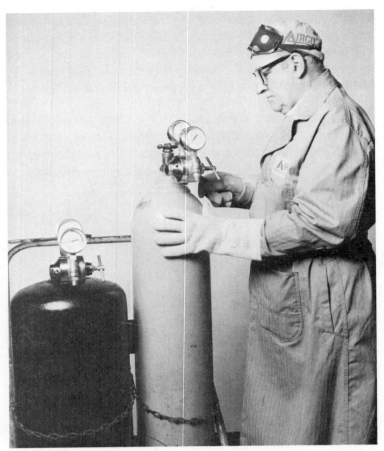

Fig. 5-3. Regulators must be attached using a minimum of force (courtesy of Airco Welding Products, Murray Hill, NJ).

the valve. Tighten the left-hand thread nut with your regulator wrench or an adjustable wrench. Sometimes the acetylene or other fuel gas cylinder will require an adapter to match the regulator threads. Get the correct adapter at your local welding supply distributor. Attach it first to the cylinder and then to the regulator. Again, make the soapy water test for leaks. Remove and clean the seats if leaks persist after moderate force is used to tighten the fittings. Turn out the pressure regulator screw on the fuel gas regulator until it, too, is loose.

CYLINDER VALVE OPERATION

There are two gauges on your regulators, whether they are two-stage or single-stage regulators (Fig. 5-7). The oxygen cylinder valve can be opened before the hoses are attached, but care must be taken that they not be opened all at once. One great turn on the cylinder valve handle may result

Fig. 5-4. Testing the connections for leakage after every hookup is the only way to be safe (courtesy of Airco Welding Products, Murray Hill, NJ).

Fig. 5-5. Back out the pressure regulator screw to keep pressure off the regulator diaphragm.

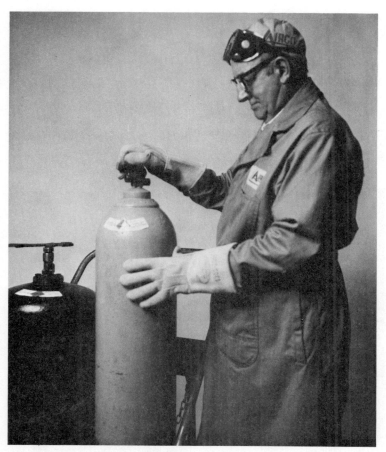

Fig. 5-6. T-handle wrenches are used to crack the acetylene valves (courtesy of Airco Welding Products, Murray Hill, NJ).

in damage to springs and diaphragms in the regulator. Open the valve just enough so you can watch the cylinder pressure gauge needle moving slowly up the face of the dial (Fig. 5-8). Once the gauge hand stops moving, you can open the valve the rest of the way. It's still easier on the gauge internals if you take your time.

You will need a T-wrench to open the acetylene cylinder valve. Start slowly and progress to the final opening only after the gauge needle stops moving. Fuel gases are handled differently. While the oxygen valve is opened all the way after

Fig. 5-7. Regulator gauges (courtesy of Airco Welding Products, Murray Hill, NJ).

the needle stops moving, the acetylene valve is opened only an additional 1½ turns. Also, the T-handle wrench is left in place on the valve so that you can quickly shut off the fuel gas if there is an emergency. Wire the T-handle wrench to the top of the tank so it won't get lost, but don't wire it so tightly that the wrench can't be turned. Leave plenty of slack.

HOSE CONNECTIONS

Oxygen hose is green, while acetylene hose is red (any fuel gas hose). Under no circumstances may they ever be interchanged for any reason. Your oxygen connectors are right-hand threads, while the fuel gas connectors are left-hand threads.

All new welding hose needs to be cleaned inside. The hose will have an accumulation of very fine *talc* on its inside, and that talc can cause some trouble if it gets inside the torch.

Fig. 5-8. Be sure you have the gauges at correct working pressures. To prevent damage, handling of the regulators and gauges should be gentle.

The oxygen hose is blown out by being connected to the oxygen regulator with one end left free (Fig. 5-9). Use a pressure of about 5 pounds per square inch (which will show on the low pressure gauge dial) to clear the hose of powder. To blow out an acetylene or other fuel gas hose, attach an extra oxygen hose connector nipple to the oxygen regulator. Twist it partway into the hose, then blow oxygen through just as you did with the oxygen hose. Never use acetylene or another fuel gas for this clearing procedure. Once mixed with air, the fuel gases need only a source of ignition to cause a fire.

Fig. 5-9. Clearing the hose of talc before attaching the torch (courtesy of Airco Welding Products, Murray Hill, NJ).

Once the fuel gas hose has been cleaned of talc, you can blow through the hose with your mouth to clear any traces of pure oxygen that might remain. This clearing process can also be carried out with compressed air for both hoses, if you have a water-filtered source available nearby.

The hoses are connected to the regulators—the oxygen regulator to the green hose and the fuel gas regulator to the red hose. Once these connections are complete, the torch can be connected to the opposite end of the hoses (Fig. 5-10).

Select the correct tip for your torch and attach it, using the procedure described by the torch's maker. Some manufac-

Fig. 5-10. Connected hoses and torch, with the welding tip being attached (courtesy of Airco Welding Products, Murray Hill, NJ).

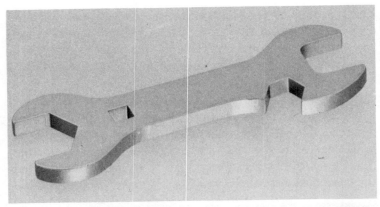

Fig. 5-11.Use only this type of wrench or a high quality wrench of the correct size (courtesy of Airco Welding Products, Murray Hill, NJ).

turers supply hand-tightened tips and welding heads, but most tips require a special wrench supplied with the outfit or available from your distributor (Fig. 5-11).

Check all seating surfaces for cleanliness and damage before going ahead with the setup. Even new tips, torch heads, and other equipment may be damaged or dirty.

LEAK TESTING

Connection tests are quite important once the hookup is completed. Mix up some soapy water and apply it on each connector after closing the torch valves and opening the oxygen and acetylene valves to working pressures (about 10 psi will do the job). Any connections that leak should be gently tightened. If leakage continues, the pressure should be cut and the hoses bled (close the tank valves and open the torch valves). Then check all seating surfaces for dirt or damage.

You can dunk the entire torch in your soapy water solution to find out if it leaks (Fig. 5-12). The torch stem packing may be at fault. First, try tightening the packing nut. If that doesn't stop the leakage, you need new packing. Use only the packing recommended by the manufacturer of your torch. Guard against oil. If the problem seems to come from a valve that doesn't close all the way, remove the assembly and use a clean rag to wipe the valve stem seats and the body clean. If this doesn't work, you need to have the valve body reseated or new parts installed.

TORCH LIGHTING

Torch lighting procedure varies, so the manual that comes with the torch should be studied. For some torches, you must open the acetylene valve all the way. For others, crack the oxygen valve and open the acetylene valve a bit. A

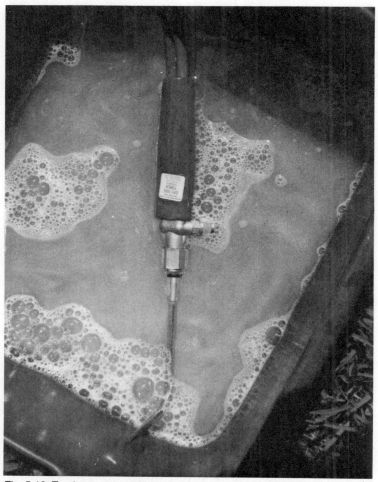

Fig. 5-12. Torches are tested for leakage by inserting the entire torch, connections and all, into soapy water.

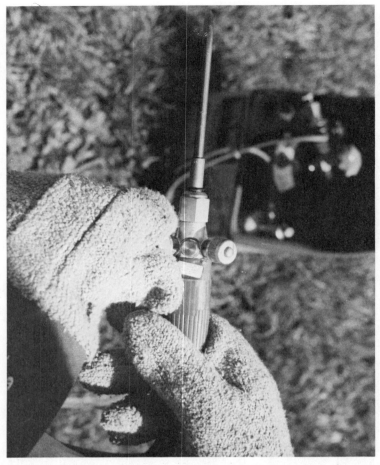

Fig. 5-13. Torch fuel valves are opened before the oxygen valves when you prepare to light your torch.

few may need no oxygen feeding at all until the fuel gas is lighted. Check first. While most torches will light with just acetylene flowing through them, it's always best to follow directions. Set the correct working pressures for both gases at the working gauges on your regulators. Open the appropriate torch valve first—usually if oxygen is required for lighting, it will be the first valve opened. The acetylene valve is opened second (Fig. 5-13). The oxygen valve will be opened only slightly in most cases. The acetylene valve opened more widely, sometimes all the way. Hold a spark lighter near the tip of the torch and light (Fig. 5-14). Remember to wear welding gloves.

The first flame you get will be an *excess acetylene* flame which burns with a smoky yellow flame, with black soot drifting off it (Fig. 5-15). The oxygen valve on the torch is now opened slowly until you begin to see the characteristic excess acetylene flame. Note the intense featheredged inner cone amid a secondary cone; the outer envelope will be blue in color (Fig. 5-16). Once the oxygen valve has been turned far enough to produce this flame (a one to one ratio of oxygen to

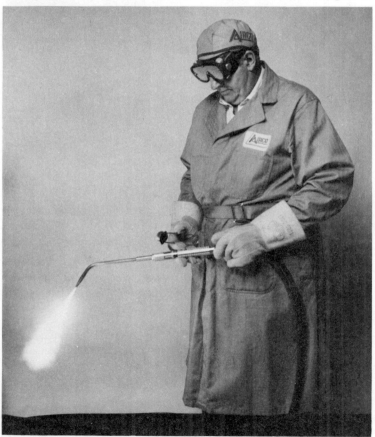

Fig. 5-14. Proper torch lighting requires a spark lighter (courtesy of Airco Welding Products, Murray Hill, NJ).

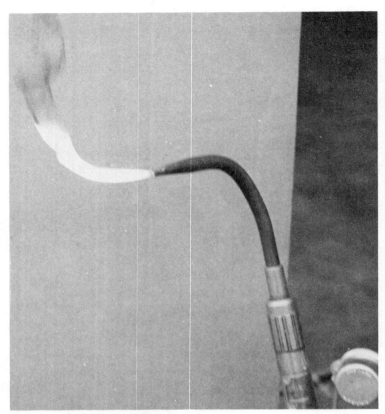

Fig. 5-15. Pure acetylene flame.

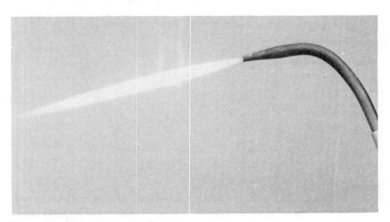

Fig. 5-16. Excess acetylene flame.

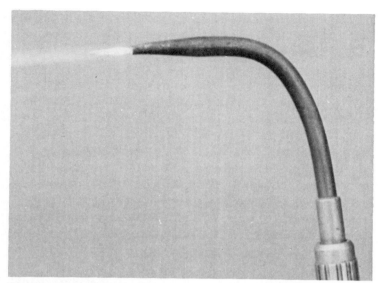

Fig. 5-17. Neutral flame (used for almost all fusion welding).

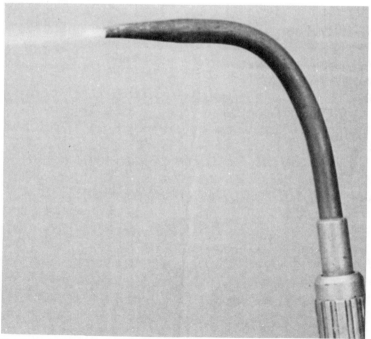

Fig. 5-18. Oxidizing flame.

fuel gas will probably still give a slightly excess fuel gas flame), it is time to adjust the acetylene valve to produce a neutral flame. The neutral flame becomes apparent as you continue slowly closing the fuel gas valve, and the secondary cone becomes smaller. Just as the cone disappears completely, you have a *neutral* flame (Fig. 5-17). If you continue reducing the fuel gas, the flame will become an excess oxygen or oxidizing flame and be smaller. It will have a slightly purple inner cone (Fig. 5-18). The changes as a flame moves to excess oxygen are harder to see even for experienced welders. At the point where the secondary cone disappears, you are where you want to be for almost all welding jobs.

You'll want to work with an excess acetylene flame in some cases. The requirement for this flame is usually expressed in terms of ratio. That is, the length of the acetylene cone (secondary cone) versus the length of the inner cone, with both being measured from the tip of the torch. If you are told to use a 2X, or two times, acetylene flame for a particular job, then the secondary cone must be twice as long as the inner cone (one X is equal length, three X is triple length, and so on) (Fig. 5-19). This type of flame, excess acetylene, is most often used on nonferrous metals. The outer envelope is the part used to do the work. In nearly all cases, the inner cone is not used to weld or braze. The heat produced is so intense that the metals will simply melt instead of being welded. Approximate flame temperature is 5500 degrees Fahrenheit (3040 degrees centigrade).

The neutral flame is used for ferrous (compounds containing iron) metals. It is the most commonly used welding flame. Again, the inner cone is not touched to the work. The neutral flame has a clear, pale blue or almost colorless outer secondary cone, and a tight, small intense white inner cone. The outer envelope will range from blue to orange. The featheredged secondary cone of the carbonizing flame becomes clear and very well defined in the neutral flame. Flame temperature will be as high as 5800 degrees Fahrenheit (3207 degrees centigrade)

The flame of most limited use in gas welding is the *oxidizing* flame. It is used primarily for welding brass and

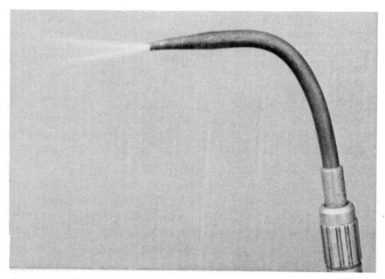

Fig. 5-19. A 2X acetylene flame.

bronze and sometimes for brazing galvanized iron. Oxidizing
flames are easily recognized as they get more intense, though
the first stages can be easily confused with a neutral flame. An
oxidizing flame has a short flame envelope, a bluish to
purplish orange secondary cone, and a short, pointed, intense
white inner cone. If the oxidizing flame is used for welding
steel, it will add oxides to the weld and cause serious strength
problems. Flame temperature is about 6300 degrees
Fahrenheit (3485 degrees centigrade).

To shut off the equipment, close the acetylene valve on
your torch. Then shut off the oxygen valve. If welding is to be
restarted almost immediately, this is all you need do. If you
are simply reclamping or replacing items to be welded, you
need only shut off the torch.

For longer periods of time, use the following procedure.
After you shut off the torch as already described, close down
the oxygen cylinder valve. Reopen the torch valve to release
pressure on the hose and regulator. Turn out the pressure
adjusting screw on the oxygen regulator so that the internal
diaphragm is not kept under pressure, causing premature
wear. Close the torch oxygen valve.

The same cylinder shutdown and hose bleeding procedure is done with the fuel gas cylinder and hose. You should never release both pressures at once, as this creates an ignition danger.

TACK WELDING

With the safety precautions and start-up procedures completed, you can begin welding. Start out with some simple practice welds, beginning with flat position tack welds in sheet metal or light metal plate using mild steel.

Sheet metal (3/32 inch or less) is difficult to weld. Its very thinness makes it extremely subject to heat distortion. For this reason, I would recommend that you locate some old auto body sheet metal and use that for most of your practice welds. If you can produce good welds, you can almost be certain of getting as good or better welds in heavier materials with only a little extra practice. Practice is best confined to mild steel where your choice of rods is wide.

Your first step is to locate some firebricks for a working surface. Under no circumstances should you ever weld on ordinary brick or stone; nor should welding be done on a concrete floor. Extreme heat expands the water molecules in these materials and cracks them. In some circumstances an explosion could result. Take a couple pieces of your practice metal, keeping the size down to about 1 foot by 6 inches each, and support them on the firebricks. Tack welds are often used as temporary supports when work is being aligned for complete seam or other kinds of welding. Where great strength isn't needed, tack welds can join two pieces of metal permanently.

Lay the metals to be joined next to each other, with firebricks supporting the ends. Light your torch and adjust it for a neutral flame. Pull your goggles down over your eyes and hold the torch loosely in your right hand—left hand if you left-handed. Grip the torch as near to its point of balance as possible. The free hanging hoses should just about balance the torch tip with the torch held loosely in your hand.

Hold the torch tip close to the spot where the joint is to be made, and apply the neutral flame to the spot where you

Fig. 5-20. As mild steel starts to turn white, it will flow and form a puddle.

97

wish to join the metal. As the metal turns red and then white, it begins to flow (Fig. 5-20). You must remove the flame from the work the instant flowing starts, or you'll burn right on through the work. To help prevent distortion, you can preheat the sheet metal by running the torch up and down the area to be tack welded a few times prior to making the welds. Making several torch passes after the tack welds are made, to help prevent rapid cooling, can also aid in preventing distortion (Fig. 5-21).

A *tack weld* is the simplest weld. It requires little time and no filler metal. As long as the alloys of the metals being joined are similar, the weld will fuse. The joint will be reasonably strong. You should continue on with your practice welds, making a series, until the process feels as simple to you as it sounds. If you've never welded before, you will find it hard to judge just when the metal is going from white hot to pudding. You will likely either remove the torch flame too quickly and get a weak weld, or you will pull the heat away too late and get a hole in the metal. Practice will allow you to tell almost exactly when the puddle will form.

FOREHAND WELDING

Once you've got the sheet metal tack welded, you can learn the *forehand* method of flat welding steel with gas equipment. Select a mild steel rod of the correct thickness (about half the thickness of the base metal to be welded). Usually any good rod meeting AWS 5.2, such as Airco's No. 7 mild steel rod, will do the job well. The forehand method of welding is the easiest to learn. When metal gets up over ⅜ inch thick, you are forced to use the backhand method since it allows greater penetration, giving a much stronger weld root.

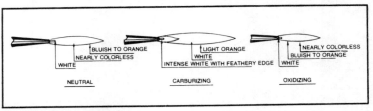

Fig. 5-21. Types of flames.

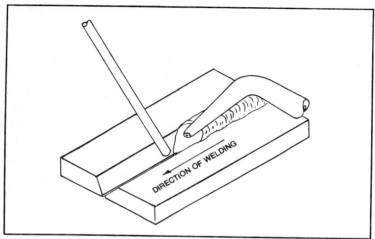

Fig. 5-22. Forehand welding (courtesy of Airco Welding Products, Murray Hill, NJ).

For most forehand welding, the torch tip will be at an angle of about 45 degrees from the work surface (for flat work) (Fig. 5-22). If the joint is fairly thick, hold your torch tip closer to the vertical. The torch tip is held near the work surface, preheating the metal and finally forming a puddle into which you place the end of the filler rod (Fig. 5-23). In forehand welding, the rod is removed from the puddle every three seconds or so, though the heated tip of the welding rod should

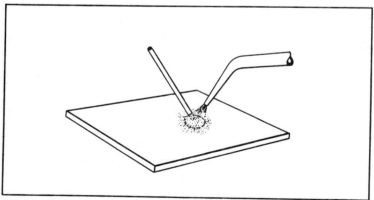

Fig. 5-23. The filler rod is placed in the torch flame (courtesy of Airco Welding Products, Murray Hill, NJ).

be kept close enough to the torch flame to keep it hot and cut down on oxidation. As your weld bead begins to form, the rod is moved on to a new puddle area.

If the torch flame and the welding rod tip are crossed in each puddle area, a slight weave effect is created. This weave is desirable and forms a strong weld (Fig. 5-24). Your torch tip should move along the weld at a constant speed. If it takes two seconds to form the first puddle, then take two seconds to form the next puddle, and so on.

Varying your torch tip speed will result in an odd looking weld—small puddles alternating with large puddles. Another result is lower weld strength because of varying penetration. While weaving the torch tip and the welding rod, make sure that the wave is small and that the welding rod and torch tip are moving in opposite directions. Practice forming a weaved bead weld until you can maintain the correct weave of both torch tip and welding rod. Keep a distance of about ⅛ inch between the work surface and the inner cone of the torch flame.

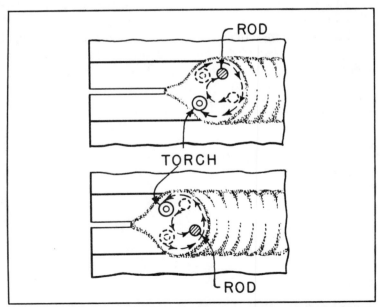

Fig. 5-24. Crossing the rod tip and the torch tip to get an even spread of weld metal (courtesy of Airco Welding Products, Murray Hill, NJ).

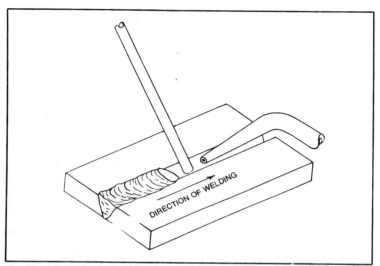

Fig. 5-25. Backhand welding (courtesy of Airco Welding Products, Murray Hill, NJ).

Preheating and delaying of cooling can again be used to help prevent extreme distortion of the thin sheet metal. The first few welds you make will distort this thin metal badly. Have plenty of practice metal around on which to work.

If forehand welding is used for thick metals, ⅜ inch and up, the metal must be grooved (flat position) so the puddle doesn't melt down into the root of the weld before that metal is ready to fuse with it. The biggest problem here is the large size of the puddle needed to fill the groove. The larger the puddle, the harder it is to control, even on flat welding surfaces. For vertical and overhead welding, control becomes impossible for the novice and extremely difficult for the professional. When thick metals must be welded in overhead and vertical positions, the backhand method of welding is desirable.

BACKHAND WELDING

Forehand welding involves moving the welding rod along in advance of the torch tip along the weld. With backhand welding, keep the torch tip in advance of the welding rod. The torch tip must precede the welding rod all along the weld, with the flame pointed back at the puddle and the welding rod (Fig. 5-25). You no longer use any weaving motion on the torch, though the welding rod may be moved in circles or arcs within the puddle (Fig. 5-26). Because smaller puddles result, control of deep flat and horizontal welds is improved. Just as in forehand welding, the tip of the torch comes closer to vertical as the thickness of the metal increases.

When using the backhand method, you will be working with metal thick enough to require a technique known as *keyholing* (the keyholing technique can also be used with the forehand method on thicker metals). Keyholing is done at the joint along the leading edge of the puddle. The root area of the weld is melted in a keyhole shape to receive the molten filler material. Backhand welding works within about the same limits of torch deflection as does welding. For most welds, keep the torch at about a 50-degree angle, and increase the angle as the work thickness increases. Only experience can determine the exact angle for each job, so practice is of great importance.

To practice backhand welding, go to an automobile junkyard and obtain some heavy pieces of steel, possibly even a portion of an old frame or an entire frame if the metal left over might be handy. If you look around the yard to make sure the chunks you select are in no shape for transfer to another automobile as parts, you should then be able to purchase on a

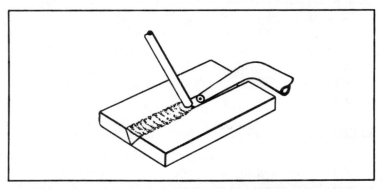

Fig. 5-26. Another example of backhand welding (courtesy of Airco Welding Products, Murray Hill, NJ).

per pound basis for about what they would bring the dealer as scrap metal.

For metal of this thickness, joint design is quite important. For practice purposes, you can grind or cut (with your torch) a 35-degree bevel about two-thirds the distance down the thickness of the metal (if it's about ⅜ inch thick where the joint is to be made). If the metal is only ¼ inch thick, you can go right ahead with a butt weld. Make sure the edge surfaces are clean and reasonably well matched.

As the metal is heated, the puddle should form and develop a fluid mirror surface before you dip the rod into the puddle. In backhand welding, the welding rod is not withdrawn from the puddle. It is moved continuously in an arc or circle, and the torch tip is moved steadily in front of it in a straight line. The flame creates a pressure that holds the puddle back until the base metal is hot enough to fuse with the filler metal. In forehand welding this pressure doesn't occur. The result can be a cold shut at the root of the joint. The puddle flows into the root before the base metal is ready to fuse, and then it closes up over the improperly fused root. The circular action of the welding rod can also be used to help hold back the puddle in backhand welding. Such a circular motion is almost essential. The rod, if it follows the torch in a straight line, has a tendency to shield one side or another of the joint, creating lopsided welds.

HORIZONTAL WELDING

With the proper amount of practice, basic gas welding techniques should soon become fairly easy for you on flat surfaces. Unfortunately, welding doesn't always get done on easy-to-handle surfaces. You need to practice horizontal, vertical, and overhead welds.

When you are doing *horizontal welding*, the forehand technique offers good bead shape, penetration, and fusion. The best procedure is to start from the left and move to the right in most cases. Hold your welding rod and your torch tip at about the same angle as you would for flat welds. Welding in this manner works well for joints in steel up to about 3/16 inch thickness. Puddle control starts to become quite important in horizontal welding. A puddle may run at least part way out of the weld, bringing on poor penetration and poor fusion. The torch is lifted from the puddle, along with the welding rod, as each part of the bead is formed. The puddle then instantly sets up and can't run when this technique is used. Then the next puddle can be started, with no worries about the preceding puddle dropping out or sagging part way down from the bead. The rod end must be kept in the torch's heat during the removal process in order to control oxidation, as is always the case with forehand gas welding techniques.

VERTICAL WELDING

The most common welding technique for *vertical welding* is the forehand method. You will usually start welding at the bottom of the weld and move from there to the top. To weld vertically, hold the torch at an angle of about 30 degrees from the work surface, with the welding rod held opposite it at about the same angle (Fig. 5-27). The pressure of the torch flame can be used to keep the puddle in place for much of the time. The bead laying procedure is the same as it is for the flat position, but puddle consistency is not. As in horizontal welding, keep your puddle from becoming free flowing as you may allow it to do on flat surfaces. The need is not as extreme in vertical welding as it is in horizontal welding because the lower portions of the bead help provide some support for the puddle. Therefore, you usually won't find it necessary to

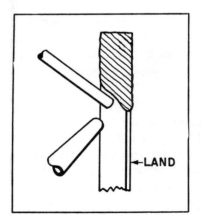

Fig. 5-27. Welding from the top down for a vertical weld requires a torch and rod position like that shown (courtesy of Airco Welding Products, Murray Hill, NJ).

←LAND

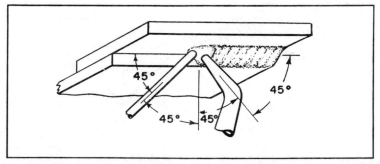

Fig. 5-28. Tip and rod angles are kept equal during overhead welding (courtesy of Airco Welding Products, Murray Hill, NJ).

move both the torch tip and the welding rod as each puddle is formed. Try your best to keep puddle fluidity down. Remember to keep your welding rod tip in the torch flame when you remove it from the puddle.

OVERHEAD WELDING

Overhead welding is the most difficult position. Simply put, molten metal, like any other liquid, does not wish to hang suspended in the air. Still, after the other welding positions are mastered, overhead welding should not require too much more time for you to finalize your competence as a welder. Again, the forehand method for most metal thicknesses is preferred. The bead is advanced *toward* the welder, though, so that flying sparks and any dropped filler metal hit the floor instead of the torch operator. Tip and rod angles differ so that the flame can be used to control the puddle flow. The torch tip angle in overhead welding can be as much as 45 degrees to the work surface and should be equaled by the welding rod angle (Fig. 5-28). Again, keep your puddle less fluid than with flat welding, and use the handy torch-welding rod removal method to keep the previous portion of the bead from dropping out of the weld. As you require greater penetration, drop your tip and rod angles closer to the vertical.

Because proper penetration with the forehand welding method requires wide V-grooves to correctly join thicker metals, the backhand method is often used for vertical, horizontal, and overhead welds as metal thickness increases (Fig.

105

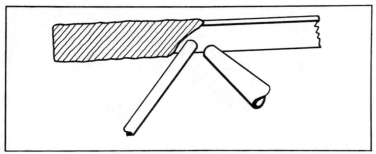

Fig. 5-29. The backhand method can also be used for overhead welding (courtesy of Airco Welding Products, Murray Hill, NJ).

5-29). This is basically because a much smaller included angle of joint is needed to get deep root penetration with the backhand method. As an example, instead of an included angle of 90 degrees or more in ⅜-inch mild steel, you would, with the backhand method, use an included angle of 60 degrees or even a bit less. In some cases, this lesser angle can actually help improve welding speed and make the actual welding fairly easier. The backhand welding position is not popular because it is hard to learn. The backhand welding method does offer automatic post heating qualities—because the torch continues to play on the already welded joint—which provide a joint of great strength and ductility.

Eyeballing a weld to see how well it has turned out should give some indication of overall weld quality. First, look for a uniform bead. Second, see that the weld is free of pinholes. Finally, look for an overall neat appearance. If you welding was done properly, so that the base metal all along the joint was melted before the filler metal was added, the resulting joint will be at least as strong as the base metals being joined.

OXYGEN FLAME CUTTING

Oxygen flame cutting is a chemical reaction, at least when ferrous metals are cut. Basically, the intense heat and the added oxygen from the cutting torch speed up the process of oxidation. The oxyfuel gas cutting torch provides an oxy-MAPP or oxyacetylene flame to preheat the metal to be cut to a point where a stream of pure oxygen can then speedily oxidize the metal along the preheat line, while blowing away much of the slag that is formed, as well as the resulting molten metal (Fig. 5-30).

Some gas cutting and welding outfits come with a set of hoses already attached directly to the cutting torch, though these are not found in prepackaged outfits. These torches have the industry standard right-hand thread nuts for oxygen and left-hand thread nuts for fuel gas, with the groove cut around the center of the fuel gas nut. The green hose is for oxygen, and the red hose is for fuel gas.

If the torch body attaches to the base, it will usually need only hand tightening, as will many of the tips used on the cutting torches. Using tips and torch bodies that require only hand tightening materially cuts down on damage to seals and threads. No oil or grease is ever used for lubricating the threads. The unit must be tested for leakage before you begin to use it. All other connections and procedures are those required when setting up the welding torch.

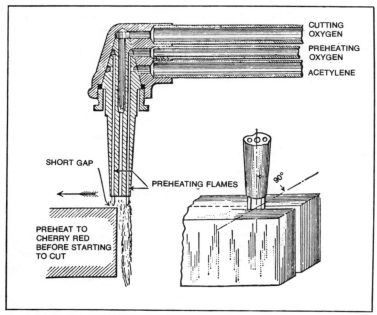

Fig. 5-30. Oxyacetylene cutting techniques (courtesy of Airco Welding Products, Murray Hill, NJ).

One major difference you'll find when attaching cutting torch tips is the need for proper positioning of the preheating orifices in relation to the type of cut to be made. On a four-hole tip, for a straight cut with perpendicular sides, the preheating holes must be aligned so that two of them follow each other along the line of the cut (Fig. 5-31). This provides a three-hole lineup across the diameter of the tip, with the pure oxygen hole in the center. To make a bevel-sided cut, you will have to position the preheat orifices so that they are set two by two. There will be two preheat orifices on each side of the central pure oxygen orifice (Fig. 5-32). This layout provides the greatest amount of heat along the lines where the metal is to be cut. Tips with more than four preheat orifices do not require positioning.

Tip installation requires that the seat on the torch be free of contamination and in good condition. The tip is then screwed in tightly. The preheat orifices are positioned. A tip tightening wrench is used to complete the job on those torches not using hand tightening only for tips.

Again, do not use excessive force when using a wrench to tighten the tips. Such force will mar the tip seat and cause poor cutting. Tips can be cleaned with the appropriate tip cleaners of the proper size.

Cutting Layout

Flame cutting is often done on sections that are flat. It can be done in any position that welding can be done, but cutting pieces laid flat will be stressed.

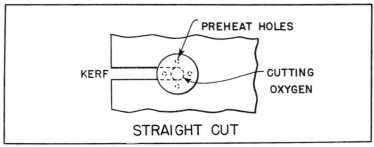

Fig. 5-31. Preheat hole alignment for a straight cut (courtesy of Airco Welding Products, Murray Hill, NJ).

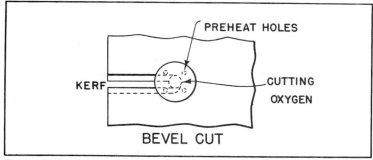

Fig. 5-32. Preheat holes set up for bevel cutting (courtesy of Airco Welding Products, Murray Hill, NJ).

If the metal needing cutting is a free piece unattached to anything else, it can be supported on firebricks. Make sure that the end being cut off has some sort of support (unless it is a small piece of scrap). Check under the cutting area so you won't be dripping molten metal on rags, wood flooring, your own gas hoses, or any other destructible material. Be sure that your cylinders are out of the cutting area, and that there are no flammable materials close enough to get splattered with molten metal. Check ventilation and for flammable liquids such as paints, kerosene, and gasoline.

You can lay out the metal with any good set of measuring and marking tools including a straightedge, chalk, and so on. You may prefer to use a tool such as Cleanweld's *Combo* square. This cutting square is made with a brass blade and three 20-pound magnets in its 45 and 90-degree frame. The larger 4-foot models have two frames, six magnets, and an aluminum blade. These Combo squares offer you two ways to get your right and 45-degree angles. You can use chalk to mark the angles with the square as a marker only, or you can simply attach the square to the work (if it's ferrous metal) with the magnets and then cut along the blades. The relatively lightweight blades do not dissolve during use. Both brass and aluminum have much better heat-dissipating qualities than steel does, so the heat flows out of the blades and frames without melting them. These squares cost more than standard carpentry steel squares. For flame cutting use, a standard carpenter's square would last less than a minute.

109

Make a second check for flammable materials under and around the material to be cut. Make sure you don't come into contact with molten metal. You can buy asbestos aprons and sleeves, gloves, and other protective items.

Take a good look at your cutting torch. The funny-looking long handle along the body of the torch is what quickly identifies it as a cutting torch or torch attachment and not a welding torch (Fig. 5-33). The valves at the base of the torch are opened. Set the preheat holes to produce a flame as close to neutral as you can get. To do this, follow a slightly different procedure than you do with a welding torch. The oxygen valve at the base of the torch is opened as wide as it can go. Then the acetylene valve is opened as on a welding torch. Your gloved hand, holding a spark lighter, will touch off the flame at the torch tip. You can then locate another valve further toward the tip of the cutting torch. This is a preheat oxygen valve. Along with the lever for cutting oxygen, it keeps any oxygen from reaching the torch tip until it is cracked, or the cutting lever is squeezed or pressed. Oxygen and acetylene (MAPP, propane) regulators are set to match the tip size and torch capacity as recommended by the manufacturer.

Adjust the fuel gas and oxygen (at the preheat oxygen valve, leaving the torch oxygen valve wide open) to get your neutral flame. Start out by opening the acetylene valve about one-sixth of a turn before snapping a spark. Increase the fuel gas flow until you have a flame jumping about 1/8 inch from the torch tip, and then back off the pressure until the flame just returns to the torch tip. Feed in the preheat oxygen easily and gradually. When a neutral flame shows, you're ready to start cutting.

Check the cutting oxygen passage by depressing the lever. Look for a dark path running through the center of your neutralizing flame. If the dark path is present, there are no restrictions. You can go ahead and readjust to a neutral preheat flame while keeping the oxygen cutting lever depressed.

Starting the Cut

You start the basic flame cut by applying the flame for preheating along the area to be cut, with the preheat flame

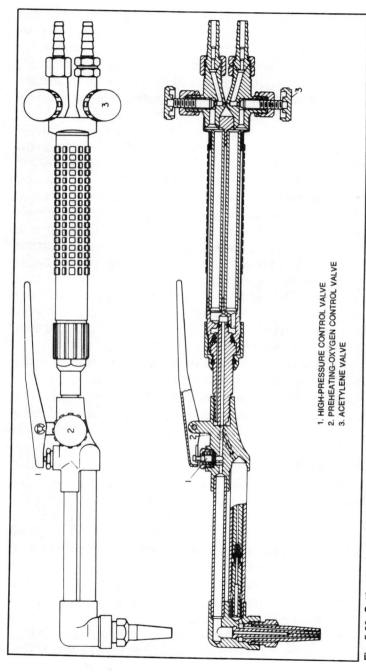

1. HIGH-PRESSURE CONTROL VALVE
2. PREHEATING-OXYGEN CONTROL VALVE
3. ACETYLENE VALVE

Fig. 5-33. Cutting torch attachments for welding torches (courtesy of Airco Welding Products, Murray Hill, NJ.

held just above the base metal surface. You are now looking for a cherry red condition in the metal being preheated. Once the cherry red color is attained, you can slowly depress the oxygen cutting lever. Cracking the lever wide open quickly could place a hard, fast stream of oxygen on the metal surface, cooling it down below the point where oxidation can take place.

Once you cut is started, you must consider your cutting speed. Cutting speed is always a variable, depending on the thickness of the metal, the size of the cutting tip, the width of the cut, and so on. You must maintain the correct cutting speed if you want a clean cut. The steadier you are in making the cut, the neater the cut is going to be. Moving too slowly through the cut will allow the cut's sides to close up with molten metal, leaving you with a really jagged cut edge. If you move too rapidly through the cut, you simply will not cut all the way through the metal.

For materials under ¼ inch thick, you should keep the torch head angle inclined at about 15 degrees from the work surface (Fig. 5-34). For work thicker than ¼ inch, hold your torch closer to the vertical.

A properly used handheld cutting torch can give you a surprisingly clean and smooth cut if care is taken. The only way to get clean cuts is to practice until you know how your torch reacts to different cut styles, depths, widths, and so on (Fig. 5-35).

Cutting Bevels

Once basic cutting is mastered, you may want to cut bevels in metal. Bevels are often needed when one metal part must fit into another, or when you are preparing to make a weld in thick sections of metal. For thick metal welding, beveling is almost essential to a good joint. Beveling is a straightforward procedure.

First, the preheat holes may need to be adjusted for bevel cutting. Your manufacturer's instructions will tell you if so. Lay out the work, mark it, and make the usual safety checks. Light your torch just as you would for straight cutting. The major difference is the angle required to get a bevel. The

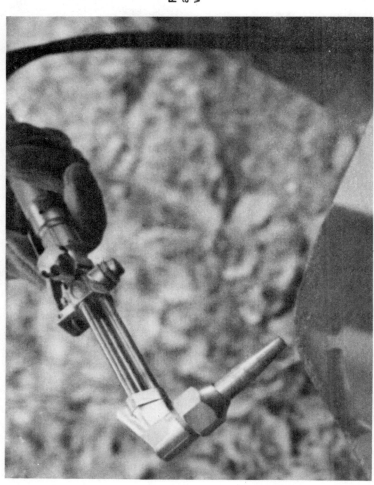

Fig. 5-34. A 15-degree cutting torch angle is needed on thin metal if you want a clean cut.

Fig. 5-35. Flame cutting with a lightweight torch (courtesy of Airco Welding Products, Murray Hill, NJ).

torch head is inclined to an angle that allows the major part of the preheating to fall on the upper parts of the bevel cut to be made, the portion where the most metal must be removed from the cut to form your bevel. The angle of the cutting torch is a function of the thickness of the metal, the angle of the bevel, and the width of the cut needed (Fig. 5-36).

Piercing and Slot Cutting

You may find a need to run a bolt or some other device through a thick slab of plate steel. You may not have the correct size drill or hole cutter, or you may not have an electric drill with the power to make the hole. Oxyfuel gas cutting comes to the rescue. Metal piercing for bolt holes and other needs is simple and serves as the start for slot cutting.

Set up as usual. Make all safety checks. Select the point where the hole is needed and mark it. Preheat the metal directly below the cutting tip—right on your mark—until it glows a bright cherry red, a bit brighter than with straight line cutting. Open the oxygen cutting level slowly. Tip the head of your torch back slightly so that sparks and slag can't bounce back up into the tip holes (Fig. 5-37). As the hole begins to form, the sparks and slag will start to drop on out the bottom of the cut. You can lower the torch to a normal cutting height as you go on to open the hole up to the size you need (Fig. 5-38).

For slot cutting, begin as if you're just making a hole in the work surface. Go on to used straight and curved line cuts (or all straight line if it is that kind of slot) to form the required size and style slot. When you are cutting a slot, the smoothest cut edges will be formed if you keep your cutting going in a continuous run. Keep practicing until you can readily form both clean perpendicular cut sides and beveled cut sides.

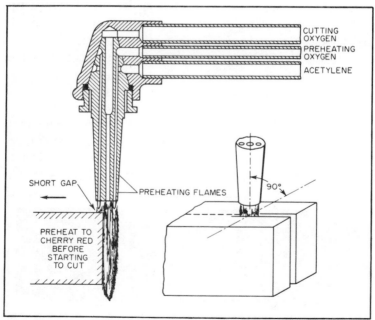

Fig. 5-36. Cutting torch beginning a cut and in action (courtesy of Airco Welding Products, Murray Hill, NJ).

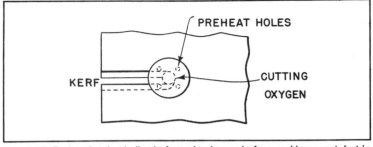

Fig. 5-37. Preheating is similar before piercing or before making a cut, but in piercing it moves directly in on the spot where the hole must be made (courtesy of Airco Welding Products, Murray Hill, NJ).

MAPP GAS CUTTING

Even though MAPP gas provides heat very close to that of acetylene, there are some essential differences in flame appearance and cutting torch angle if you want optimum performance. First, you need to change tips. Your MAPP gas cutting tips will have slightly larger preheat holes than will those you use with acetylene. If you use an acetylene tip (which you can do safely), efficiency will suffer a slight drop. You should never use tips for natural gas or gas other than acetylene with MAPP gas. These tips are not manufactured to withstand the heat produced by either acetylene or MAPP gas when used as a fuel gas with oxygen. The tips will overheat dangerously, which is a prime cause of flashback.

Torch lighting and flame adjustment, along with the rest of the startup procedure, are so similar to acetylene that no further comment is needed. The flames themselves will have slightly different appearances. The MAPP gas flame is yellow

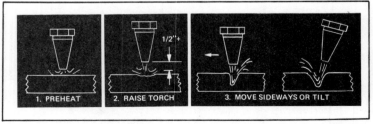

Fig. 5-38. As soon as the hole is pierced, the torch tip must be drawn back to normal cutting height.

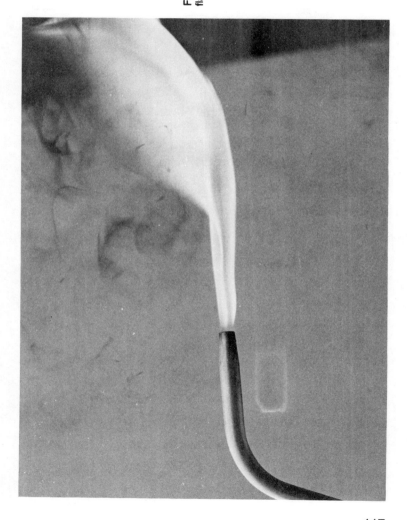

Fig. 5-39. Extreme excess MAPP flame.

117

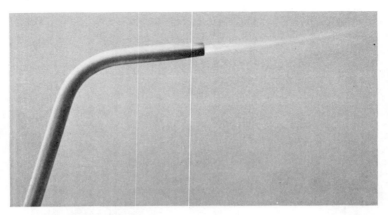

Fig. 5-40. Excess MAPP flame.

and, as oxygen is fed in, becomes blue with a yellow, feathered secondary cone inside the heat envelope (Fig. 5-39). The tip cone on this carburizing flame is a dark blue (Fig. 5-40). As more oxygen is added, the yellow secondary cone will disappear. You will see a dark blue tip cone inside the heat envelope when you have a neutral flame (Fig. 5-41). As you obtain an oxidizing flame by adding more oxygen pressure, you will get a harsh hissing sound. The tip cone will turn a lighter blue (Fig. 5-42).

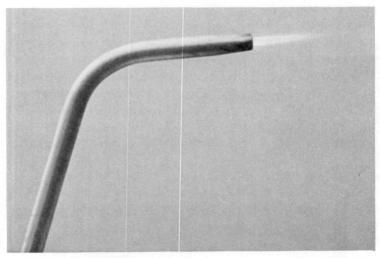

Fig. 5-41. Neutral MAPP flame.

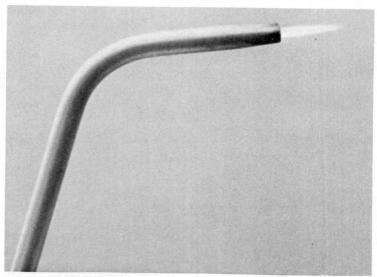

Fig. 5-42. Oxidizing MAPP flame.

When using MAPP gas, try not to get a flame identical to that of acetylene in appearance. If you get a flame looking like an acetylene flame would, the preheat it provides will not be great enough to allow cutting. The MAPP gas flame you finally get will be at least 1½ times the length of a similar acetylene flame (Fig. 5-43). Don't try to reduce the size of the flame to the acetylene length.

The oxygen cutting jet on a MAPP gas cutting torch is known as the *stinger*. If the stinger is correct, the cut will go along as quickly and possible a bit more smoothly than would an acetylene cut. If the stinger is not correct—if the volume of the oxygen is too low, or the pressure is too low or too high—you will have problems with the cut. Slag may stick in the bottom of the cut line as you move along. The slag sticking may be caused by too much oxygen pressure, which can blow out the slag before oxidation is complete (oxidation is what makes the cut). The same condition may be caused by a cutting speed that is too fast. If you have slag sticking at the bottom of your cut line, first try slowing up your cutting speed. If slowing up doesn't improve things, reduce your oxygen pressure a bit and try again.

119

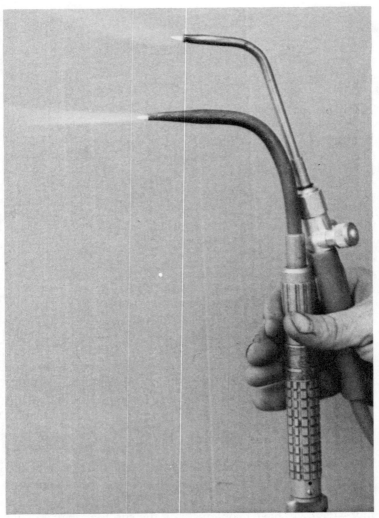

Fig. 5-43. MAPP-acetylene flame comparison. The top flame is MAPP and is oxidizing, while the bottom flame is neutral acetylene.

If the bottom of the cut line opens up, your oxygen pressure may be too low. Increase oxygen pressure a bit, or pick up your cutting speed a little. Check the cut line again.

As with acetylene, cutting light metal requires that you tilt your torch head somewhat to prevent distortion of the metal from excessive preheating. Use an angle of about 45 degrees from the surface of the metal being cut. Keep the tilt along the line of the cut (Fig. 5-44). If the metal to be cut is less than ½ inch thick, you should increase the torch tilt to 65 degrees. You don't need to do this with acetylene since metal of that thickness requires a vertical torch. Using MAPP gas, you use a vertical torch only after the metal is over ½ inch thick.

While the speed you make on a cut will vary with the individual jobs and depends on tip size, metal type, and metal thickness, some indications that you are holding the correct speed can prove valuable, especially when you're first starting to make cuts. Look at the spark stream flowing from the cut. If it shoots out in the direction the cut is being made, then you're holding the correct cutting speed. If the spark stream is flowing straight down or possibly even coming back toward

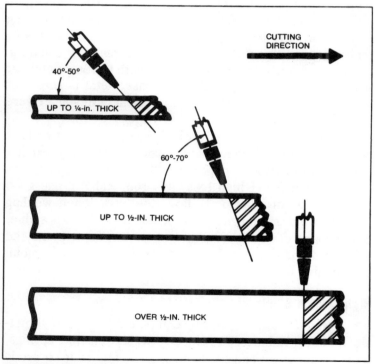

Fig. 5-44. MAPP cutting requires a tilt of 45 degrees at the tip on thinner metals (courtesy of Airco Welding Products, Murray Hill, NJ).

the torch tip, you are then moving too rapidly through the cut. Slow things down if you want a really smooth cut (Fig. 5-45).

WELDING DIFFERENT METALS

Most of what I've already covered for practice welding will apply to virtually every form of mild steel you're likely to need. Unfortunately, all the metals you're going to find it necessary to work with will not be mild steel. Not all will be ferrous metals, so there are a few basic things to check out beyond accurate welding rod selection.

Wrought Iron

The melting point of *wrought iron* is about 2750 degrees Fahrenheit (1511 degrees centigrade). Because of this high melting temperature, wrought iron requires more heat to weld or cut than does mild steel. Wrought iron gives more indication of different composition since the slag melts a long time before the base metal goes molten. This early slag melting gives the metal a greasy appearance, a condition you'll become used to when working with ferrous metals. It will often indicate to you that the base metal is ready for welding; here it does not. Apply more heat until you are sure the base metal is melting.

When welding wrought iron, careful rod selection is of great importance. A check with a manufacturer's representative or your distributor might be helpful. Linde recommends their Oxweld No. 1 high test steel rod, and I've had good results with Airco's No. 1 alloy steel rod. When welding wrought iron, try to agitate the puddle as little as possible. Begin your weld by first concentrating your flame at the end of the rod, then change that concentration to the base metal surrounding the puddle.

High Carbon Steels

If you don't have the shop equipment to heat treat metals, it is best to have someone else weld items like springs, tool steels, *cutlery*, and so on. Each type of metal is correctly heat treated for the job it does, and correct heat treating without the proper equipment is extremely difficult.

Fig. 5-45. MAPP cut in progress.

Cast Steels

Most of the cast ferrous metal welding you do will be on cast irons. Occasionally some part made of cast steel, used where greater ductility and shock resistance are needed, will snap or otherwise break. Most often such a break will occur because of some casting fault, such as a sand hole. On larger cast steel parts you have to preheat the parts to prevent distortion, and preheating is usually a good idea for even small parts. Look for indications of distortion; cast steel will bend more than cast iron will before breaking. Major distortion of the casting is possible. Do any heating and bending needed to restore the pieces to their original shape before you begin welding.

Welds in cast steel should be beveled, using either a cutting torch or a grinder. Weld with mild steel rod. Carefully preheat thick sections until they are a bright red. If you notice bright spots or craters along the weld, there are spots where the casting was faulty. Work any sand or slag to the surface of the puddle, using the torch flame pressure, and then move the slag out as close to the edge of the puddle as you can. On any cast steel piece where preheating was necessary, you get the best results by surrounding the welded piece with insulating material to prevent too rapid cooling.

Chromium and Nickel Chrome Steels

Chromium and *nickel chrome steels* are stainless steels. Because of the difficulty in identifying the three types, and the poor results that occur if some types are not correctly treated during welding, I would strongly recommend that you use braze welding or hard soldering to form items or make repairs with these metals.

Braze Welding

Gas *braze welding* is one of the handiest metal joining techniques any do-it-yourself welder can learn. Braze welding is a bit easier than fusion welding. It doesn't provide quite as strong a joint, though it provides a joint far stronger than any style of soldering. Braze welding uses joints designed like those for fusion welding, but there is no fusion of base and filler metals. Braze welding allows lower joining temperatures, so that heat distortion problems are considerably lessened. Also it allows you to join dissimilar metals with a strong joint.

Generally, braze welding rod melts around 1600 degrees Fahrenheit (872 degrees centigrade), well below the melting point of most commercial metal alloys. This relatively low melting temperature allows you to join many metals that would otherwise have to be soldered or bolted together. One additional process that also involves braze welding is known as *bronze surfacing*. Bronze surfacing is a method by which worn surfaces are built up.

When braze welding is done, the molten metal from a bronze welding rod spreads over a surface (base material) that has been properly treated in a process called *tinning*. When the base metal cools and the bronze filler becomes solid, the bond formed between the two gives excellent strength, though they are not fused.

In order to get proper tinning action, you must carefully prepare the joint. Basically, the joint to be braze welded is prepared just as are all other joints to be welded. Thicker metal must be beveled. There are a few special joint designs that are used specifically with braze welding in order to increase joint strength. Joint cleanliness is of much greater importance for braze welding than fusion welding, as temper-

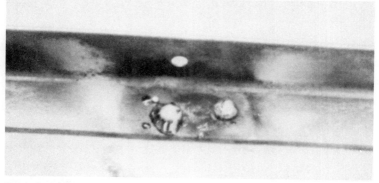

Fig. 5-46. If the temperature of the base metal is not high enough, the brazing rod will form balls instead of flowing correctly.

125

atures almost never get high enough to burn off oxides. All rust, scale, grease, oil, paint, and other foreign material must be removed. You need a suitable flux to remove oxides that will remain on the surface after mechanical cleaning is complete. Apply braze welding flux by dipping the end of the rod into powdered flux before applying it to the base metals and torch flame.

Proper base metal temperature is reached when the metal begins to glow. If you apply the rod before the welding temperature is high enough, the molten bronze will not flow over the surface of the base metals. It will form drops or balls and run down to the bottom of the joint (Fig. 5-46). If the base metal is too hot, you will get little balls instead of drops, and the bronze will boil. If the temperature is correct, the bronze will flow evenly and spread over the surface you're working on with no problem (Fig. 5-47). It will take you two or three tries to recognize the correct flow, but after that the job is simple.

If braze welds must be built up, make sure that the second and any succeeding pass is thoroughly bonded to the first. Such bonding should prove no real problem.

Cast iron is one of many metals that is quite hard to weld, along with stainless steels. Cast iron is hard to weld because of the oxides in the base metal which tend to be drawn into the joint, while stainless steel can give a poor joint because of granular changes in the base metal. Both metals can be braze welded easily. The braze welding of cast iron will often

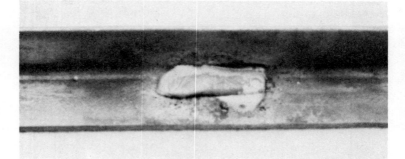

Fig. 5-47. Both silver solder and brazing rod will flow very well when the temperature of the base metal is correct.

produce a stronger joint than would be possible with fusion welding. Small particles of graphite appear to melt out of the cast iron structure. The molten bronze then flows in, or seems to, and increases bond strength. Braze welding uses a filler material, bronze, that slowly yields to stresses as it cools. Internal stresses in the cast iron may be displaced, creating a stronger whole.

Obviously, though, braze welding is not for materials that are going to undergo temperatures higher than the melting point of the filler metal. You should also remember that the braze welded joint will lose most of its strength at a temperature of no more than 500 degrees Fahrenheit (260 degrees centigrade).

Braze welding is not useful on steel under stress. This is seldom much of a problem, though, since stressed steels can usually be fusion welded fairly easily.

SILVER BRAZING

Silver brazing is also known as hard soldering and silver soldering. It is an oxyfuel gas process that can often be accomplished using one of the new single gas torches with MAPP gas as a fuel. Lighter jobs can be carried out with some of the newly designed propane torches which improve gas and air mixing and increase flame temperature. You can often do light braze welding jobs with these torches. Strength of the joints made with silver soldering falls somewhere between braze welding and soft soldering. If the alloy layer or tinned area is kept as thin as possible, and your joint design is good, tensile strengths on the order of 40,000 psi are possible (though seldom reached). In silver soldering there is very little buildup of material in the joint being formed, for the solder flows through capillary action. This capillary flow requires a certain degree of clearance, or it will be impeded. Your joint clearance distance can't be too great, or the buildup of the relatively weak filler metal will be too thick, thus greatly reducing joint strength. Joint clearances of .003 to .006 inches will provide the optimum strength for silver brazed joints. Such joints withstand vibration very well, making this form of soldering an excellent choice for repairs to stereo speakers and air conditioners.

Corrosion resistance of silver brazed joints is also exceptionally high. Silver brazing techniques are handy when assembling such items as photo lab sinks of stainless steel and other such equipment exposed often to strong acids or other chemicals.

Clean joints are essential to good silver soldering. The lower temperatures of the process cannot possibly burn off grime and oxides. A fine grade of *steel wool* or *emery cloth* can be used for mechanical cleaning (Fig. 5-48). Flux is used for chemical cleaning and is most often applied with a brush to both the joint and solder or braze filler material. A neutral flame is used to heat the base metal surfaces of the joint, and the surfaces in turn melt the brazing alloy. The alloy itself is then touched to the base metal. If the preheating is sufficient, it will flow over the fluxed parts of the base metals. Because the filler alloy will cover all fluxed parts, you should not get the flux on sections where you don't want coverage. The torch flame is never touched directly to your solder or brazing alloy.

Silver soldering can now be carried out with equipment less costly than oxyfuel gas gear. Almost all jobs can be handled with a MAPP gas torch which provides more than sufficient heat to melt the filler alloy, though for the very largest jobs you might want to go back to oxyfuel gas. A propane torch in some of the new torch designs will hand smaller silver brazing jobs (Fig. 5-49).

HEAT TREATING

Heat treating is found in various forms today, with a few types being of interest to do-it-yourself welders. You will have to go to a metals fabricator or a distributor for information on the temperatures and times needed to carry out many heat treating projects.

Quenching—the dipping of hot metal in water, air, or oil—is of little primary importance in welding processes covered here. *Hardening* requires finding the critical range of the metal being treated, heating it to about 100 degrees

Fig. 5-48. Emery or crocus cloth can be used for the mechanical cleaning of copper pipe prior to sweat soldering.

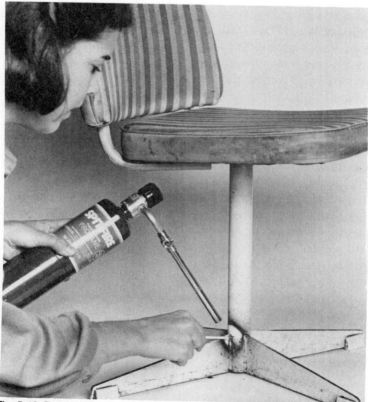

Fig. 5-49. For silver brazing, a single gas MAPP torch is often used instead of oxyfuel gas (courtesy of Airco Welding Products, Murray Hill, NJ).

Fahrenheit (38 degrees centigrade) above that point, and then quenching it. This can be done as a complete hardening, or a case hardening where the outside layer of the metal is hardened and the inside left soft (a method frequently used with chains and locks).

Annealing is a process designed to soften steel. You must heat the steel to 100 degrees Fahrenheit (38 degrees centigrade) above its critical range, and then hold the temperature constant for about one hour for each inch of thickness being softened. Slow heat buildup is recommended. Once a magnet drops off the surface of the metal being treated, the steel is at its critical point (it loses its magnetic qualities). For total annealing, a sealed container would be needed.

Normalized steel is harder than annealed steel but still retains a great degree of ductility. Some high carbon steels will become so hard with this process that they can not be machined. Gas welders will be able to normalize only small items, and most are best done in a furnace. The process requires you to heat the steel a few more degrees than you would for annealing. The steel is then allowed to return to normal air temperature with no further interference.

Flame hardening is likely to be useful to the gas welder. If an item such as a lawn mower blade has been repaired, you can take it, after all machining operations, including final sharpening, have been completed, and heat the blade edges to just a bit above the critical point. Then you quench the metal. Air quenching will provide just about the same results as described for normalizing. Water does a good job of quenching low alloy steels. *Brine*, a 10 percent salt solution with water, will be less likely to cause interior cracking of the metal than a plain water quench will, though.

When the time comes to improve the durability of shovels, mower blades, plow tips, and other metal parts subject to abrasion, special methods and rods are easily found. A few special procedures are sometimes needed to get the best results from these welding rods.

HARD SURFACING

Hard surfacing means that a welding rod material is

bonded to the edge of a tool where that edge is subjected to excessive friction, abrasion, temperature, or impact. The special alloy of the hard surfacing rod will cut down wear on new tools. It can also be used to build up new work surfaces on old tools, helping to cut down time and replacement costs.

According to Eutectic Corporation, their tungsten carbide *TungTec 10112* welding rod gives a Rockwell hardness of 57 to 64. TungTec 10112 may prolong service life of some tools by as much as 400 percent. Abrasion resistance and friction resistance are tops. There is an advantage of hard surfacing softer steels to make them more suitable for extreme wear conditions. The hard surfacing can be repeated each time the coating does wear out until the base metal finally fatigues to the extent that it can no longer be used. Hard surfacing allows the use of more ductile, softer metals.

You must first select the hard surfacing rod that has the qualities to match the job that you want the hard surfaced tool to carry out. Take the tool to be hard surfaced and clean the surface thoroughly by using a grinder, wire brush, or other suitable tool. All sharp corners should be rounded off, but don't grind down a needed sharp edge. Make sure there is absolutely no rust or scale left on the surface.

Set up for preheating, though in most applications the thickness of the metal won't be great enough to require extreme preheating. Look for a dull red heat (you'll have to use goggles with clear lenses). You must keep the work area fairly dim to see it. You can get a heat indicating pencil that changes at about 800 degrees Fahrenheit (427 degrees centigrade). Make sure the surface is uniformly heated. Cooling distortion may crack the hard surfacing alloy and, in some cases, cause it to lose its bond with the base metal.

Adjust your torch to provide a slight carburizing flame (slight excess acetylene or MAPP gas flame). Adjust the fuel gas feather for the particular rod being used. Eutectic recommends a ½ to 1X carburizing flame with their Durotec 19910 rod, while Linde's Haynes Stellite Alloy No. 92 requires a 2X flame. Others may require as much as a 3X flame. Under no circumstances should you allow your flame to become oxidizing. This will destroy the hard surfacing process and might ruin the surface of the base metal.

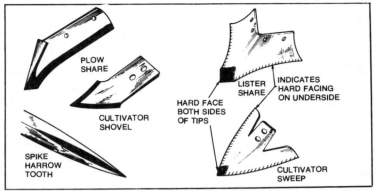

Fig. 5-50. Hard surfacing patterns for tools that must be self-sharpening (courtesy of Sears, Roebuck and Company).

Either backhand or forehand torch styles can be used to apply hard surfacing rod. You should use a tip about one size larger than that used for welding with the same diameter rod. Keep the inner cone of the flame about ⅛ inch off the work surface, and use a 15 to 20-degree torch tip angle. For extremely thin edge hard surfacing, tilt your torch until it is more nearly 60 degrees. As the thickness of the metal increases, you can bring your torch closer to the vertical (all torch angles are from the vertical). Different patterns of bead are required for different tool uses, but in most cases you won't find it necessary to be fancy. Remember that a tool which gets its major abrasion on one side, and little or no abrasion on the other side, will last almost as long if the side getting minor abrasion is left untreated. Hard surfacing only one side of a tool this way will make the tool almost self-sharpening. As the softer metal wears away, the sharp edge will remain (Fig. 5-50).

Oxy-MAPP or oxyacetylene hard surfacing is best for most thin-bladed tools. There is little chance of burning up the working edge of a shovel with the ease of heat control you have and the relatively low bonding temperatures of the hard surfacing alloys now available.

Hard surfacing slag can sometimes be very difficult to remove. If you can't get it off with wire brushing, try special grinding wheels.

Chapter 6
Joint and Weld Design

Occasional welders are more likely to make mistakes when laying a bead, so that the proper joint design is probably even more important than it is for the professional welder. Joints have different degrees of strength. You must apply the most useful and strongest joint design, along with the strongest weld configuration possible within the limits of work design and space. In some cases one type of strength will be needed, while in others you'll have to try and gain another type of weld strength.

In some cases *shear resistance*—the ability to resist tearing or ripping forces—will be of utmost importance. In others, vibration resistance will be more important. You may want to combine the two forms of resistance, while also adding strength in some way to keep impact or shock loadings from destroying the weld.

It isn't within the scope of this book to get into metal stress loading computations. You will need to apply your common sense when determining the primary type of loading likely to be applied to a weld in a particular situation.

In most automobile welds, vibration will be a minor factor. Sometimes it will be a major factor. Items such as trailer hitches will also have to resist some strong shear stresses trying to tear the welds apart. For other jobs like new shock mountings, you must allow for plenty of shear strength, a fair amount of vibration, and then consider possible stresses. *Shock stresses,* or *impact loadings,* become really special problems for those vehicles used off-road where a boulder may bounce off almost any part of the undercarriage at any time or speed.

Generally, working most of the time with mild steel, you will find that good welds in properly sized materials will resist all stress loadings that don't deform the base metal. Exotic materials are seldom necessary. Essentially, your job is to design the weld/joint combination for maximum penetration and fusion to allow weld strength to equal the strength of the original metal.

JOINT DESIGNS

There are five basic welding joint designs: *butt, corner, edge, lap,* and *tee* (Fig. 6-1). These joint designs can be divided further into several subtypes or combined. More than 90 percent of your jobs can be accomplished with one of the basic joint designs. The job dictates any combinations or adaptations necessary.

The butt joint appears to be the simplest weld. A square butt joint has the surface planes of the base metal welded or fused. Ways to add strength include *beveling, V-jointing,* and *U-grooving.* A butt joint welded from both sides of the base metal is known as a *double joint.*

In corner joints, the pieces to be welded form an L-shape, most often with an angle of about 90 degrees between the pieces of base metal. Neither piece will extend beyond the other, but a range of lapping styles can be used. With corner to corner lapping, the edges of the joint barely touch (Fig. 6-2). A half lap corner joint has one piece overlapping the other piece by half the thickness of the metal. A full lap joint means that the pieces are brought up edge to edge. Corner joints can be beveled, U-jointed, or veed, from a single side or from both sides (Fig. 6-3).

Edge welds require *flanges* or at least one flange. To make an edge joint, you will have to fuse the outer surface planes of the base metal pieces. Single and double flanges are the most common edge joint styles. Again, edge joints can be

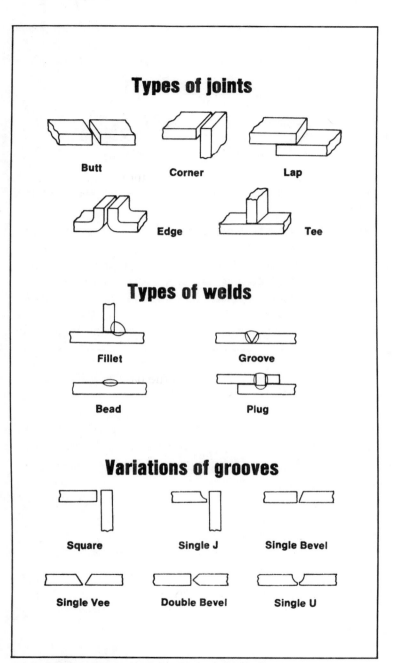

Fig. 6-1. Types of joints and welds (courtesy of Hobart Brothers Company).

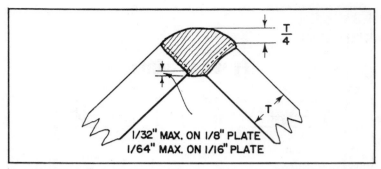

Fig. 6-2. One style of welded corner joint (courtesy of Airco Welding Products, Murray Hill, NJ).

beveled, veed, or U-jointed, though none are used with any real frequency.

Lap joints are made by overlapping two pieces of metal and welding at the point where they overlap. Lap joints can be made with a *double fillet weld* or a *single fillet weld* (Fig. 6-4).

Tee joints are made when one piece of metal, as in the corner joint, meets another at an angle of about 90 degrees,

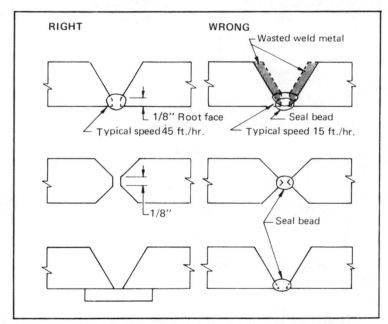

Fig. 6-3. Groove variations (courtesy of Hobart Brothers Company).

forming a shape like the letter T. Tee joints differ from corner joints because the tee joint always has an overlap from one piece of base metal. Beveling, veeing, and U-jointing can be used on one or both sides of the base of the tee joint.

Once you know what the basic joint styles are and can easily visualize each one, you should have no trouble coming up with any needed variations on the five basic styles. Any edge joint can be held at angles other than 90 degrees, as can tee joints. They are no longer basic joints when this change of angle takes place.

JOINT QUALITIES

Assigning a weld joint design to a particular job is easier if you realize some of the qualities of the various designs. Each has good and bad points regarding time consumed, filler metal amounts needed, and strength.

Butt joints tend to use more filler metal than single lap joints do. They can require a lot of preparation time if the metal is thick enough to require grooving or beveling. Butt joints will do an excellent job when you need a weld that will withstand bending or shock stresses. They also make the best welds for containers that hold flammable materials. In thicker metals, getting sufficient penetration for good fusion of the base metal and filler metal at the root of the weld requires special preparation.

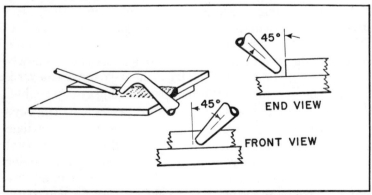

Fig. 6-4. Lap joint single fillet weld (courtesy of Airco Welding Products, Murray Hill, NJ).

Lap welds are among the least used weld joints for several reasons. First, when you use a single lap weld, joint resistance to bending stresses is very low. Making a double lap weld requires twice as much time, and more rod or electrode, than an equally strong simple butt weld does. Lap welds are good on cylinders and are sometimes used on containers that carry flammable materials. The simple butt weld is preferable. The butt weld is stronger than the single lap weld, faster and cheaper than the double lap weld, and safer than either lap weld. A lap weld may allow some of the flammable material, especially on a double lap weld, to penetrate the area between the two welds and become trapped there. Such entrapment means that the container cannot be cleaned and made safe for repairs or other uses, since you would always face the dangers of contamination or explosion.

Corner joints, tee joints, and flange joints are used where needed. Welds and weld designs are used to add any strength necessary to any joint.

WELDING POSITIONS

Welding positions have already been covered in the chapter on gas welding, and they will be included again in the chapter on arc welding. A slightly different look is needed because the position in which an article must be welded can do much to determine the type of weld joint and weld bead you need to use. Flat welding is always to be preferred whenever its use is possible. Puddle control, weld consistency, and strength are much better when welding can be carried out flat.

Vertical welding is usually begun at the bottom of the joint, with the lower part of the bead helping to support newly molten filler metal and base metal as the weld moves upward. It is a good second choice when flat welding isn't possible. Horizontal welding adds a few more problems than does vertical welding, as the puddle is not supported by a previous bead or puddle that is at least partially solidified. In horizontal welds the puddle will try to drop out of the joint if the welder doesn't take care to keep the puddle plastic instead of fully molten. This need for plasticity makes it hard for you to get complete penetration and good fusion on horizontal welds.

Overhead welding is the most difficult because the forces that make horizontal welding a problem are even more extreme (Fig. 6-5).

WELD TYPES

Once you've decided on an appropriate joint design, you will need to figure out which type of weld, either *groove* or *fillet*, is best for your purposes. The groove weld is basically a butt joint weld, as it is made in the groove between two pieces of metal. Fillet welds are made on other joint designs such as corner, tee, and flange. Thus, the fillet weld joins two pieces of metal at approximate right angles to each other, while the groove weld joins two butted pieces of metal.

There are a number of standard styles of groove welds, at least in part because of the popularity of butt welding as a single side welding process. A double square groove weld means you have welded both sides of the groove between two pieces of metal.

A simple vee groove weld means that one side of the butt joint has been beveled to allow greater penetration. A double bevel groove means that you have beveled both sides of *one* piece of metal and then welded from both sides.

A single vee groove weld means that both sides of the groove—both pieces of the base metal—have been beveled before welding is carried out. A double vee groove weld means that both sides of both pieces of base metal have been beveled, and the weld is always made from both sides.

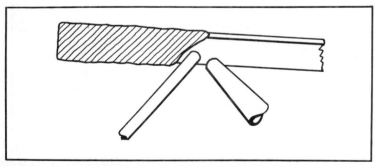

Fig. 6-5. Make a root pass on an overhead butt weld (courtesy of Airco Welding Products, Murray Hill, NJ).

A single *J-groove* weld is one in which a shape like the letter J is ground into the edge of one plate of base metal. This provides for excellent filler metal penetration while also leaving some support for the molten root in thicker stock. The double J-groove is in one piece of metal only, leaving a lip in the center as a root support for those times when welding is done from both sides of the base metal.

Single U-groove welding means that both pieces of base metal have been ground in a J-shape providing, when they are placed together for welding, a U-groove. A double U-groove means that both sides of both pieces have been ground and welded.

Fillet welds don't offer the variety you find with groove welds. Basically, there are only two—the *convex fillet weld* and the *concave fillet weld.* A full fillet weld is a weld with its size equal to the thickness of the thinner of the two pieces being joined. A partial fillet weld, obviously, is not equal to the size of either of the pieces being joined (Fig. 6-6).

The simple tack weld has already been described in the chapter on gas welding. Tack welds with gas equipment are made without using filler metal. They are most often used to hold parts in correct alignment for final welding. Almost all electric arc welding will require that you use an electrode to provide filler metal for tack welds, though with TIG equip-

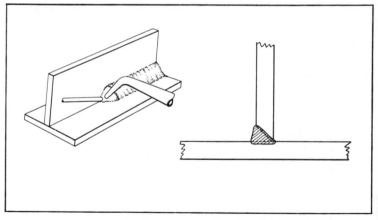

Fig. 6-6. End view of a fillet weld (courtesy of Airco Welding Products, Murray Hill, NJ).

ment and carbon arc torches tack welds can be made with arc welding equipment using no filler metal.

A *surfacing weld* is the type of weld you'll be using if you decide to hard surface some tools or other implements. The surfacing weld normally consists of one or more string or weave beads made over the area being surfaced.

Plug welds are usually circular. They are used to join one member to another above (or below) it, through a hole in one piece of base metal.

A *strap weld* is basically a butt weld. A *backing strip* has been welded to the back or bottom side of the groove to be welded (Fig. 6-7).

In addition to beveling or otherwise grinding or cutting the joint to shape in preparation for welding, you should clean the metal's surface of contaminants before starting to weld. If your flame or arc is not hot enough to remove most contaminants, they will cause joint weakness, leading to early failure of the weld. Usually wire brushing or power grinding will be sufficient to prepare the surface for welding, but with some metals you will need to use a cleaner especially made for prewelding cleaning. This is true when you're welding copper, aluminum, and some other fast oxidizing metals with oxides that are very difficult to remove by mechanical (grinding or wire brushing) means.

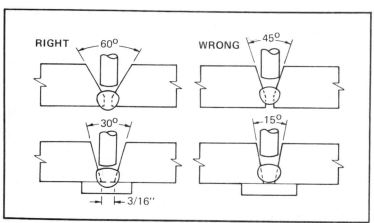

Fig. 6-7. Backing strips help to assure good root penetration in large, overhead, or otherwise difficult welds (courtesy of Hobart Brothers Company).

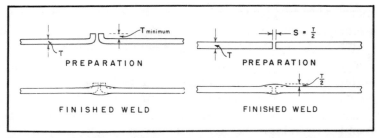

Fig. 6-8. Preparations for different types of welds (courtesy of Airco Welding Products, Murray Hill, NJ).

BEVEL GROOVE AND U-GROOVE JOINTS

If the metals you're working with are less than ¼ inch thick, there is seldom any need for you to use any weld joint design other than flat surface butt welds. Still, it's best to weld metals from ⅛ inch thick to ¼ inch on both sides if beveling isn't done. This assures good penetration and fusion at the weld root. These light thicknesses also require what the experts classify as an adequate root opening. The pieces of base metal should be separated by about 1/16 inch along the weld joint (Fig. 6-8).

For metals that are from ¼ inch to ½ inch in thickness, or even ¾ inch thick, you will need at least a single bevel groove joint. Single J-groove joints are also suitable for metals up to ¾ inch thick. They can be used easily with most butt joints, tee joints, and corner joints. Double U-groove joints are suitable for metals up to 1½ inches thick if welding is done butt style.

For beveled, J-grooved, and U-grooved welds, there is a need for specific or near specific angles at the sides of the preparations (for the sides shaped to vees, bevels, and the like). These angles help to assure proper penetration and maximum joint strength.

A single vee butt joint will need about a 60-degree angle, while a double vee butt joint will need only a 45-degree included angle. As the metal thickens, you need to open up the angle of the joint to get good penetration and root fusion. Especially with arc welders, you may have to open up the angle of the groove in order to allow the electrode entrance to the weld at the correct angle.

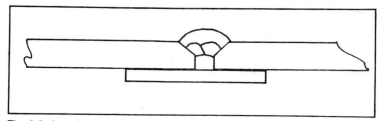

Fig. 6-9. A backing strip on a weld.

Backing strips can be used so you do not have to weld moderately thick materials from both sides. A backing strip will allow you to weld one side only of materials ⅜ inch, or a bit more, if the backing strip is allowed to be welded in and to become part of the base material (Fig. 6-9).

WELD BEADS

Weld beads come in many shapes and forms. Use little or no side to side motion of the torch and rod, or the electrode, when making a gas weld stringer bead. The type of welding bead used for materials over ¾ inch thick is called a string bead, and it follows the requirements for all groove welding. Most gas welding is carried out as bead welding, though the torch can also be used to weave weld.

There are so many variations on weave welding beads that it's impossible to cover them all. Some types of weave welding beads build up filler metal quickly, while others can be used to add design to the parts being joined. Figure 6-10

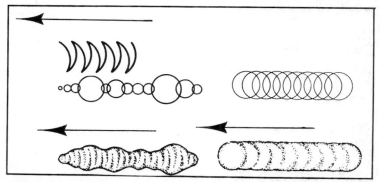

Fig. 6-10. Weave weld patterns (courtesy of Airco Welding Products, Murray Hill, NJ).

143

shows a few types of weave welding beads including *zigzag,* *oscillating,* and *circular. Pause points,* which aid fusion and prevent undercutting of the sides of the weld, are used in all weave welds. Pause points always fall at the outer edges of the weld.

Chapter 7
Arc Welding Equipment

The do-it-yourself welder has many styles of arc welding equipment from which to choose. You can select a 50-ampere lightweight model and use it on a standard 110-volt, 20-ampere circuit. You can use a 295 or 300-ampere machine for heavier-duty welding and cutting. You can select an ac/dc machine which will provide great versatility. Many companies now make tungsten inert gas attachments. Some can be used only with ac/dc machines. Others are used with the high frequency units required before ac machines can go to TIG work. Airco makes a metal inert gas unit that is expensive. Portable arc welders are available at reasonable prices.

You probably have little need for the larger welding equipment such as Airco's 450-ampere model CV-45 (MA) or the 1500-ampere automatic welders offered by Hobart and Lincoln Electric. These companies offer smaller units that are suitable for the do-it-yourself welder.

POWER SOURCES

For every arc welding job you do, you must have enough current (amperage), of the correct voltage, to maintain your arc. Current may be either alternating or direct, but you must be able to control it with precision. The welder or power source does this job. Power sources are classified as to the type of current (ac or dc, or ac/dc) and the amperage output. The voltage output is often included in the classification. Also included is the classification for the method in which the power source is fed its power, either from a power line or from some sort of engine-driven generator.

The ac arc welders are transformer machines. Direct current welders may be *transformer-rectifier* or *motor-generator* units. In the rectifier welders, the rectifier converts ac to dc, while in the motor-generator welders the motor or engine drives a generator that puts out dc. An alternator is needed to produce ac. Gasoline and diesel engine-driven welders are made with all types of current output.

Ratings are set by arc welder manufacturers. They must be in accordance with standards set by the *National Electrical Manufacturers Association* (NEMA). If the welder you want to buy does not meet NEMA specifications, look for another. The machines are rated according to the amount of current they are capable of putting out at a specified rated voltage over a *duty cycle* expressed in percentage. This duty cycle is the percentage of a 10-minute period that the welder can safely be kept in operation, without overheating, at a given current setting (in amperes). As an example, consider a 300-ampere machine rated at 60 percent. This welder can be operated at a 300-ampere setting for six out of every 10 minutes with no danger of overheating. To operate the welder 100 percent of the time, you should reduce amperage until current is reduced to 230 amperes or less. In almost all cases, machines for home welders will be rated from 20 to 60 percent. A 60 percent setting is ample for virtually all farm and home welding.

Duty cycle ratings are probably far more important for professional welders than for do-it-yourself welders. Most people seldom have the welder set at maximum amperage, meaning that the need for using 100 percent capacity, or running time, is available much of the time. Automatic, wire-fed welders are much more likely to run full speed all the time. There is little harm in waiting for the welder to cool down every 10 minutes or so.

TRANSFORMER ARC WELDERS

A voltage step-down transformer is used to change high voltage (115 to 230 volts), low current (20 to 60 amperes) electricity to low voltage (varies with the machine), high current (up to about 300 amperes) electricity for use with ac arc welding. In the United States and Canada this ac will cycle, or swap polarities, 60 times each second. This means that the voltage drops through zero 120 times a second, creating an unstable arc condition. The design of the ac arc welder and the particular electrodes are today used to surmount any problems brought on by this unstable arc.

Transformer arc welders are compact and can be fairly inexpensive. It's possible to get a 230-ampere ac arc welder, with rated duty cycles from 20 to 100 percent, delivered to your home for about $200, though local and state taxes and shipping costs may raise that figure. If you've already got a 230-volt, 60-ampere circuit, the only other cost to you will be some electrodes. The welder comes with cables, electrode holder, welding helmet, and ground clamp. The unit is the Montgomery Ward Power-Kraft 5995R, but it is by no means a unique welder. Lincoln, Sears, and many other companies offer almost identical machines.

For your lighter welding needs, both Sears and Montgomery Ward offer 50-ampere arc welders that will operate on standard 115-volt circuitry (using a 20-ampere time delay fuse or circuit breaker). These lightweight arc welders can be set up for well under $100 (Fig. 7-1). Because of cost and compactness, ac arc welders have become very popular for the home and farm, as well as for shop maintenance welding. The reasons were good originally and remain just as sensible today (Fig. 7-2).

RECTIFIER-TRANSFORMER WELDERS

Most of the direct current rectifier-transformer welders in use today are using *silicon* rectifiers. Essentially, these are transformer welders with a rectifier added to convert the output from ac to dc. Those welders available to the do-it-yourselfer are of the variable voltage type, just as are the transformer welders for ac use only. Those welders you are most interested in provide the capacity for ac welding as well as for both dc polarities. The welders are versatile. Horizontal and overhead welding can be carried out with direct cur-

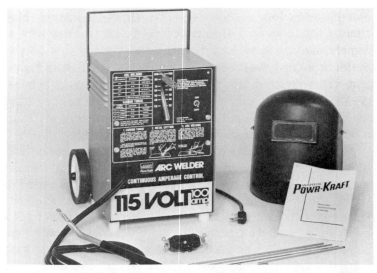

Fig. 7-1. Lightweight, low amperage arc welders can often be plugged directly into existing circuits in top shape (courtesy of Montgomery Ward).

rent, while alternating current is used for flat position, vertical welding, and variety of other jobs. Added to this versatility is the ease of converting a dc welder to tungsten inert gas arc welding for use with metals that are difficult to weld like aluminum. The ac welders cannot be converted to TIG welding use without the addition of a fairly costly high frequency unit in addition to the TIG conversion kit. The dc welders require only that you use the argon shielding gas conversion. They do not require the extra cash outlay for a high frequency arc stabilizer. Even though the ac-dc arc welder costs a bit more originally, the TIG conversion kit saves more than that amount. A TIG conversion outfit including the gas hoses, gauge, regulator, torch, and so on will, when added to the cost of a high frequency arc stabilizer, just about double the price of your arc welder. The high frequency unit costs about half that amount. You can see the savings involved in buying ac-dc equipment if you ever plan to go to TIG equipment.

Tungsten inert gas welders can be used for any welding procedure that can be carried out with nongas shield ac or dc arc welders, with a limit of about ¼ inch thickness for the weld. The TIG units are the handiest when you have to work with hard-to-weld metals like chromium, nickel chrome steels, and aluminum. TIG welding can also be done at extremely low amperages—often as low as 10 amperes. The shielding gas and the lower arc temperatures help to keep the temperature of the base metal down, cutting down on both

Fig. 7-2. Lincoln Electric Company's 225-ampere ac welder (courtesy of Lincoln Electric Company).

changes in the internal metal structure and on general heat distortion (Fig. 7-3).

Motor-generator welders are not likely to be seen in do-it-yourselfers' workshops. These are large, unwieldy machines that are quite expensive. The few units you may use are those in the lower price ranges produced by companies like Lincoln Electric and Belco. These units are used by people who need portability and who cannot get by with any kind of gas welding equipment on their jobs. (Fig. 7-4).

SELECTING WELDERS

Proper equipment selection depends in large part on knowing what jobs you're most likely to undertake. Look at the equipment needed to carry out each procedure on various metals. Notice metallic objects around your home that may need repair. You will have a good idea of what equipment you must have right away and what can be bought later.

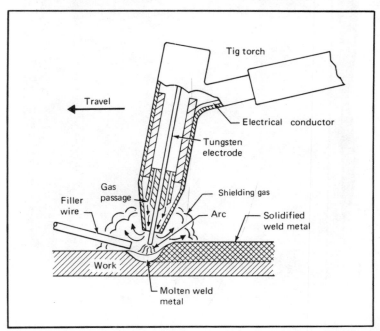

Fig. 7-3. Principles of the tungsten inert gas welding process (courtesy of Lincoln Electric Company).

Fig. 7-4. Lincoln's portable 150-ampere ac welding machine. This welder is handy for welding away from a workshop (courtesy of Lincoln Electric Company).

The size of the machine and its rated output in terms of amperes, along with duty cycle ratings, is of importance. The higher the amperage rating, the thicker the metal you will be able to weld. The duty rating needs to be considered if you expect to be doing a lot of maximum power welding (at the top amperage ratings). Operating above the rated capacity of the welder occasionally will probably not harm the welder, but daily operation over capacity will blow circuit breakers and eventually ruin the welder. Most manufacturers of arc welders adhere to a solid safety factor to prevent harm from inadvertent or intermittent overloads.

Consider cost. If you're not likely to be doing overhead or horizontal welding or you don't expect to ever need a TIG conversion, then a dc welder can be a waste of money.

Lincoln Electric Company provides a power source guide to arc welder selection, from which the following ideas apply to do-it-yourself welders.

□ **Dc Preferred Operations.** Fast freeze applications; fast follow applications; welding stainless steel; nonferrous electrode use; surfacing with high alloy electrodes.

□ **Ac Preferred Operations.** Fast fill applications; iron powder electrode use; prevention of arc blow.

□ **Ac or Dc.** Depending on the job requirements, gas tungsten arc welding.

The preferred power source for most flat welding on plate or other heavy metals is ac because of the easy usability and low cost. If you expect to do sheet metal work, dc is preferred because of its greater delicacy. If you will do overhead, horizontal, and flat welding, the machine should be a combination ac/dc welder.

WELDING CABLE

Welding cable *must* be matched to the capacity of the welding machine. The cable from the power source to the workpiece, attached by a heavy clamp, is the *ground cable*. The cable from the power source to the electrode holder is the *working cable*. It is used to strike the arc and complete the circuit used for welding. The cables are of equal importance, as no arc can exist without a complete circuit. Both cables are needed for that circuit. In most small arc welders, the cables will be supplied in the correct size. They will not be so heavy that your ease of movement is impeded. If the arc welder is powerful enough to require extremely stiff cable, a *whip section* will be used as a working cable. This whip section is reduced in size, to the minimum permissible diameter for its own length, from the main cable (Fig. 7-5).

ELECTRODE HOLDERS

At the end of the working cable you will find your *electrode holder* (Fig. 7-6). The electrode holder consists of a clamp with replaceable jaws, an insulated handle, and a connector, either soldered or bolted on, that holds it on the working cable and provides electrical contact. Never use an uninsulated electrode holder or one with badly abraded or worn insulation. Your arc welder, if new, will come with an electrode holder designed to fit the most common electrode sizes that your machine can handle. If you have to select a new electrode holder for any reason, pick one for the smallest size electrodes you commonly use. The smaller the electrode holder, the better it fits in a normal sized hand. Your working control is improved, and your hand tires less easily. If any

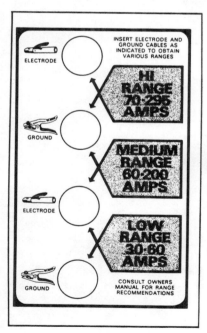

Fig. 7-5. Arc welder operating controls (courtesy of Sears, Roebuck and Company).

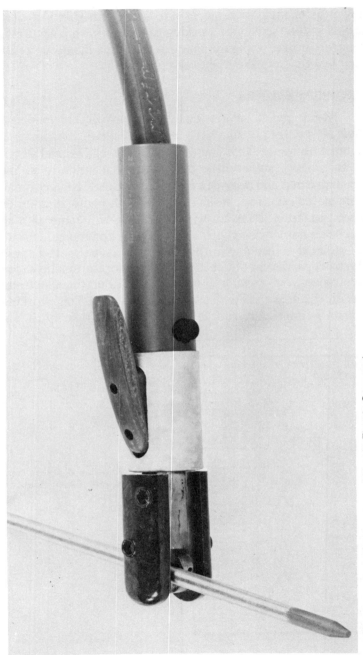

Fig. 7-6. Electrode holder (courtesy of Lincoln Electric Company).

electrode holder heats up when in use, replace it with the next larger size. Generally, electrode holders are rated from 300 to 600 amperes. The do-it-yourself welder will usually select the 300-ampere model.

GROUND CLAMPS

To form any basic electrical circuit and allow current flow, you must create a good ground. Your arc welder's ground clamp will be bolted to one end of the ground cable. It is attached directly to the work in most cases. Place your ground clamp as far as possible from the area being welded, cut, or brazed. This reduces your chances of welding the clamp to the work surface.

Because a good ground is essential to any good arc welding job, you should check the ground clamp and cable to make certain that the spring is still strong, the teeth are getting a good bite for good contact, and that the cable is not badly cracked, frayed, worn, or damaged (Fig. 7-7). When you attach the ground clamp to the work surface, use a twisting motion to make certain it bites through any residual grime or rust to get down to the bare metal. Use sandpaper or a wire brush if you are unable to get a good, full jaw clamp contact with the base metal in any other manner.

OTHER ARC WELDING EQUIPMENT

When you use the stick electrode welding process (manual shielded arc and manual arc welding are other names for the same process), your electrode is consumed and provides filler metal to complete the bead and joint. When you begin to

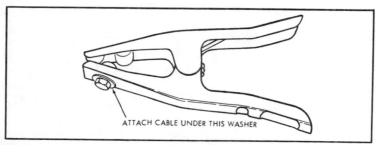

ATTACH CABLE UNDER THIS WASHER

Fig. 7-7. A ground clamp (courtesy of Sears, Roebuck and Company).

155

use tungsten inert gas arc welding procedures, the tungsten electrode is not consumed or, rather, is classified as nonconsumable because of its very slow rate of erosion. With TIG welding, you will feed in a stick of filler metal to complete the weld (Fig. 7-8). An inert gas, probably either *argon* or *helium,* is fed to surround the arc and prevent airborne contamination of the weld (oxide formation is cut drastically). Equipment to convert to TIG welding (allowing rapid switchback to straight ac or ac/dc) will about double the cost of a top quality ac model, plus the cost of the gas and cylinder.

The *carbon arc torch* is another useful accessory (Fig. 7-9). The carbon electrodes are consumed, but at an extremely slow pace when compared to standard filler electrodes. Carbon arc welding, using filler metal rods fed in from the side opposite the torch, looks very similar to TIG welding, though no shielding gas is used (the rod may be fluxed). Carbon arc welding is a procedure seldom used today because stick electrode welding gives better results in almost all cases. Still, if an arc welder is your only heat source, you'll find the carbon arc torch quite handy for braze welding and, at very low settings, hard soldering. The carbon arc torch will produce a better weld in aluminum and various copper alloys.

Both tungsten inert gas welding and carbon arc welding use an open arc to provide heat on the work surface. The work

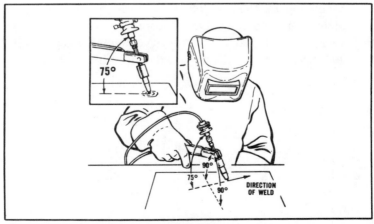

Fig. 7-8. TIG welding (courtesy of Sears, Roebuck and Company).

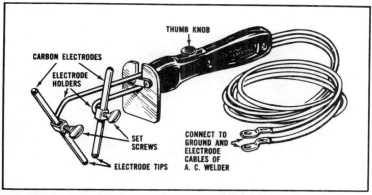

Fig. 7-9. Carbon arc torch of the twin electrode style (courtesy of Sears, Roebuck and Company).

surface is independent of the arc. While this independence is important in extending the basic uses of these two types of welding, it also allows the use of *plasma arc* welding. Essentially, this type of arc welding passes the arc struck by a nonconsumable electrode through a constricting hole in the presence of a plasma producing gas. The plasma arc is the result. A plasma arc is directional to a very high degree. It provides exceptionally deep weld penetration while creating a very small heat affected zone.

PROTECTIVE EQUIPMENT

The protective equipment needed for arc welding is similar to that needed for gas welding. Gloves, arm protectors, and aprons are important. Long-sleeved shirts are needed to prevent arc radiation burns on the arms. A welding helmet is worn in place of goggles for eye and face protection. Never look at an electric arc with the naked eye.

INSTALLING AN ARC WELDER

If a suitable 230-240 volt, 60-ampere circuit isn't in place already, your best bet is to select a safe area in your shop away from easily combustible walls, floors, and ceilings. Have a professional electrician run any required new circuitry. If your house or shop wiring hasn't been updated for some time, you'll almost certainly find that an extra service panel,

Fig. 7-10. A fire extinguisher should be nearby when welding.

or even a full new entrance panel, will have to be installed to carry the loads. Entrance panels are commonly loaded beyond rated capacity, but there are strict limits to how far this over loading can go. It is probable that no 100-ampere main entrance panel will handle the circuits needed for arc welding. Even if your house or shop wiring is reasonably up-to-date, new circuitry may be needed. Most residential 230-240 volt circuits are set for loads of 30, 40, or 50 amperes. Most arc welders require you to work with nothing less than a 60-ampere circuit.

The best area for arc welder installation will be away from the main shop (in terms of safety, not necessarily efficiency) where the walls, floors, and ceiling have all been fireproofed. Such a location isn't often possible; then you must take other precautions. Walls or floors must be covered with fire-resistant substances. All welding must be done over asbestos or firebricks. Several inches of sand in a metal pan can also make a good floor surface. Any combustible materials must be removed, while flammable walls are protected with flameproof curtains or materials like bricks. Keep a good fire extinguisher on hand. Make sure the extinguisher is rated BC and is of the dry chemical type (grease and flammable liquids cause type B fires, and electrical fires are type C). Most extinguishers are also rated for type A fires (wood, paper, cloth, and so on). You should never use any extinguisher classified only for type A fires. These extinguishers use water, and electrocution can result when they are used on electrical fires (Fig. 7-10).

Chapter 8
Arc Welding Procedures

Arc welding may seem more difficult than gas welding. Actually, though, modern electrode design and the equipment you can obtain have made arc welding easier.

After electrode and heat range selection, probably the hardest part of arc welding is the judgment needed to decide on the correct welding speed for each job. Such judgment comes only with time, practice, and application of one or two principles I will discuss shortly.

After you've chosen the correct alloy electrode and determined that it is the correct size for the job, select the required heat range. Use the lowest heat range possible. The thicker the weld needed, the more amperage you must use. From that point, the job starts with striking your arc. Again, as with gas welding, locate some metal about ¼ inch thick and about 6 inches by 1 foot in dimension. Support the ends of the metal on firebricks. Hook up your arc welder with the clamp going on the base metal work surface. Practice striking an arc smoothly.

Place the electrode in the electrode holder and select the correct heat range (Fig. 8-1). Choose an electrode compounded for mild steel—the best material for you to use for practice (and probably the cheapest)—and about half the thickness of the metal on which you will be striking the arc. The manufacturer of the electrode will have specified the heat

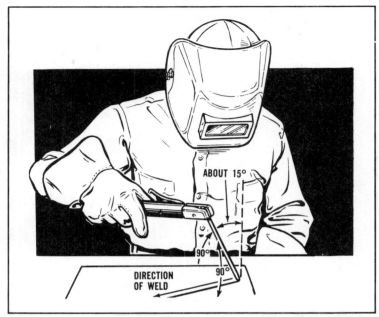

Fig. 8-1. The correct electrode angle is set before pulling down the face shield on the helmet (courtesy of Sears, Roebuck and Company).

ranges for the electrode you're using. Start on the low side. For example, perhaps you select Lincoln Electric's 6011 electrode for ac use, which is the Fleetweld 35. With a ⅛-inch electrode, using ac, your amperage range is 80 to 130. My recommendation would be to start with 80 amperes. For a beginning welder, it is best to begin with a definite end point so it is easier to tell if anything is wrong. As you gain experience, you can start in mid-range or wherever you want and still tell almost immediately what is wrong with the arc, or the bead, and what changes are needed.

STRIKING ARCS

Tilt your electrode about 15 degrees from the vertical, after setting the arc welder to 80 amperes. Strike the tip of the electrode on the work surface as you would strike a match; make a scratching motion along the metal surface (be sure your helmet is in place before attempting to strike the arc) (Fig. 8-2).

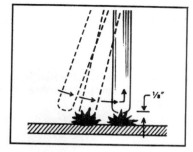

Fig. 8-2. Striking the arc (courtesy of Sears, Roebuck and Company).

As the arc is struck, the darkness inside your helmet will disappear. If you place the electrode within ½ inch or 1 inch of where you wish to strike the arc before lowering your helmet, it will be easier for you to strike the arc in the correct spot. As your arc lights off, you raise it from the work surface about ¼ inch. Finally, bring the arc back down to the work surface until it is about the diameter of the electrode from the work surface. Lowering the arc in this manner provides you with the short arc that gives good penetration and very little spatter. Your downward movement to get the short arc is made just as soon as you see a small, molten puddle begin to form.

This arc striking technique should be practiced until you have about covered your first practice piece of metal with strikes. As you continue with the practice, sooner or later the problem known as arc freeze or sticking will occur. When sticking occurs, you have touched the electrode to the work surface after the arc is struck, causing a short circuit and welding the electrode to the base metal. Give the electrode holder a firm twist, and you will usually find that the frozen electrode loosens (Fig. 8-3). If the electrode doesn't snap loose quickly, release it from your electrode holder. Allowing it to remain could overheat the electrode and cause damage to the welder. If you were forced to release the electrode, shut down the welder. Twist the stuck electrode off the base metal with a pair of pliers.

Striking an arc is a bit different if you decide to use a dc welder. You must select a rod to correspond with the AWS E6010 rating (assuming you're working with mild steel). Generally, you will use reverse polarity. To strike your dc

163

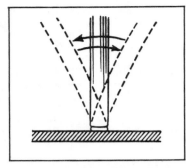

Fig. 8-3. Releasing a frozen rod (courtesy of Sears, Roebuck and Company).

electrode arc, you simply tap the tip of the electrode lightly on the work surface.

RUNNING BEADS

Once you can strike an arc consistently and well, the next step is to run a bead. Once your arc is struck, the electrode starts to be consumed. The base metal becomes molten and fuses with the electrode's filler alloy. Since arc temperatures may go as high as 6000 degrees Fahrenheit (3318 degrees centigrade), fusion is extremely rapid in thin metals such as those you will be using for practice. As the electrode melts, you must move its tip closer to the work surface to maintain the correct short arc distance, so you continue to get good penetration and low splatter. If at the same time you move the electrode along the surface of the metal, you get a weld bead or a string of molten electrode fused with the base metal (Fig. 8-4). To run a bead, the electrode almost is held straight up and down, with an angle from the perpendicular of no more than 15 degrees. This 15-degree angle is maintained in the direction of the electrode's travel. Once your puddle is twice the width of the electrode's diameter, you can begin to move the electrode along the bead. This is a help, too, in judging correct weld speed. Practice in running straight weld beads is needed so that you can develop smoothness in lowering your electrode which, in turn, results in a smooth and strong bead (Fig. 8-5).

Running a series of beads in the same direction across or along your piece of practice metal is a good idea. If the *pockmarks* from arc striking practice interfere, simply flip the

Fig. 8-4. Practice stringer beads (courtesy of Sears, Roebuck and Company).

piece over. While running your practice bead, aim for smoothness and consistency of bead width while running your arc along the surface of the metal (Fig. 8-6).

Remove electrodes from the electrode holder when they get down to about the last 1½ inches. Sooner or later you will come up with a finished electrode in the middle of running a bead, or at least before the weld is complete. You need to start a new arc. Begin the new arc about 1½ inches *ahead* of where you broke off the weld to change electrodes. Carry your bead back to the original weld to cover all striking marks, and then continue on with the bead (Fig. 8-7).

Learning to tell when you're making a good bead is a problem for many people, but here are a few tips which can help. To check for penetration, simply lift the electrode from the bead breaking the arc immediately. The crater you see formed will show the depth of penetration that you've been

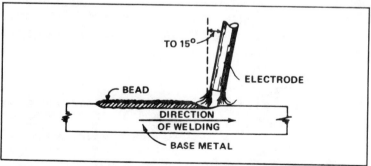

Fig. 8-5. Electrode angle for running a string bead on flat work is about 15 degrees from the vertical (courtesy of Airco Welding Products, Murray Hill, NJ).

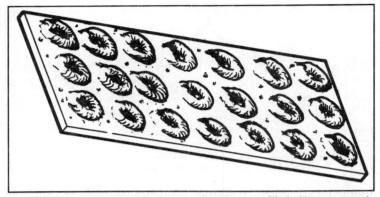

Fig. 8-6. By the time both sides of the practice piece are filled with craters made by striking arcs, you should have the knack of getting started (courtesy of Hobart Brothers Company).

getting. If your bead is uneven, several things may be wrong, though rate of travel is the most likely. Keep a smooth, consistent pace, and your bead will be smooth and consistent, with a center that is slightly above the edges and of a rounded shape. If the bead is both flat and uneven, you are probably holding too long an arc. Long arcs diffuse their heat giving poor penetration, greater amounts of spatter, and a poor looking weld bead. If penetration is poor, the amperage setting on your welder may be too low. If the sides of the bead show signs of a condition known as *undercutting* (the base metal is melted away from the bead edges), then your amperage setting is too high. Spatter also increases with too high amperage. Too slow electrode movement creates a piling effect, giving you a bead that is high, wide, and lumpy in looks. If your base metal is quite thin, for example sheet metal under ⅛ inch thick, you might get a combination of effects from too

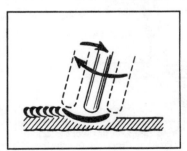

Fig. 8-7. Starting a new bead, after changing electrodes or stopping for any other reason, requires moving to ½ inch in front of the old bead, restriking the arc, and welding back to the old bead before continuing on (courtesy of Sears, Roebuck and Company).

166

slow electrode movement, with a piled bead combined with holes in the base metal where the electrode burned through (Fig. 8-8).

You should run beads in practice until you are getting a consistent result, with each bead forming a good looking weld. You will also learn to identify a rather crisp sound like frying bacon that comes with correct travel speed and arc length. You'll be running a bead about twice the width of your ⅛-inch diameter electrode, with an arc just as long as the electrode is thick.

As you come to the edge of your base metal plate, a crater tends to form at the end of the weld bead. These craters must be filled to form a good weld bead. Your filling process is a simple one. Run the arc to the edge of the plate, stop for a moment, and finally reverse the travel of your electrode for just enough distance to complete the filling of the crater.

Practicing with the electrodes you expect to use most often is a good idea because each electrode type has, at best, slightly different reactions to arc length, speed of travel, and so on. As an example, E6012 and E6013 electrodes, while sensitive to arc length, are not nearly as sensitive as E6010 electrodes, which are more likely to stick to the work surface if your arc length is too short.

If holding arc length is one of your major worries, some electrodes can help you alleviate this problem. If a manufacturer says an electrode is especially suitable to use with a drag technique, as Lincoln Electric does with their version of E6013 (Fleetweld 57), you have almost no worries about holding arc length. Simply strike an arc and drag the electrode along the work surface, at the recommended angle, on its coating. The coating holds the arc length correctly.

WEAVE TECHNIQUES

Once you've reached this point, arc welding improvements are a matter of practicing a variety of welds and weld bead techniques in several positions. Some more information on the use of weave techniques may be helpful.

Weave techniques are essential when the width of the groove you must fill, or the size of the fillet, is greater than

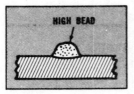

CURRENT TOO LOW
Arc is difficult to maintain. Very little penetration. High bead.

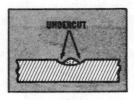

TRAVEL TOO FAST
Small bead undercut in some places. Rough top and little penetration.

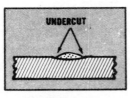

CURRENT TOO HIGH
Wide thin bead, undercut. Crater pointed and long. Rod burns off very fast.

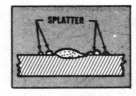

ARC TOO LONG
Surface of weld rough. Rod melts off in globules. Arc makes hissing sound.

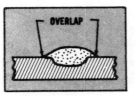

TRAVEL TOO SLOW
Metal piles up, making a wide heavy bead, over-lapped at sides in places.

NORMAL CONDITIONS
Uniform ripples on surface of weld. Arc makes steady crackling sound.

Fig. 8-8. Welding problems (courtesy of Sears, Roebuck and Company).

the available electrode diameter. You could run a series of straight beads, one over the other, but a properly done weave bead produces a better and neater weld more quickly.

If you come upon a weld, though, that exceeds five times the electrode diameter, you still need to use multiple passes, even using the weave technique. You can cut welding time greatly using weave beads for multiple passes. A straight or stringer bead will be barely twice the diameter of the electrode (Fig. 8-9).

Each weave technique requires even more concentration than stringer beads. In every weave design optimum spots where you'll pause for an instant with the electrode in order to get maximum penetration. The most frequent pause areas are at the edges of the weaves, with movement pretty continuous throughout the actual center of the weave motion, whether your weave is an arc with sharp edges or a series of arcs with squared or arrowed edges (Fig. 8-10).

If your electrode runs out in the middle of a weave, then you must follow almost exactly the same procedure as you would when running a straight bead. Let the weave cool down

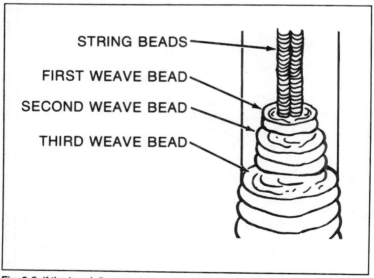

STRING BEADS

FIRST WEAVE BEAD

SECOND WEAVE BEAD

THIRD WEAVE BEAD

Fig. 8-9. If the bead diameter is over five times the width of the electrode being used, then weave techniques will not suffice for single pass work. You need to use multiple passes (courtesy of Hobart Brothers Company).

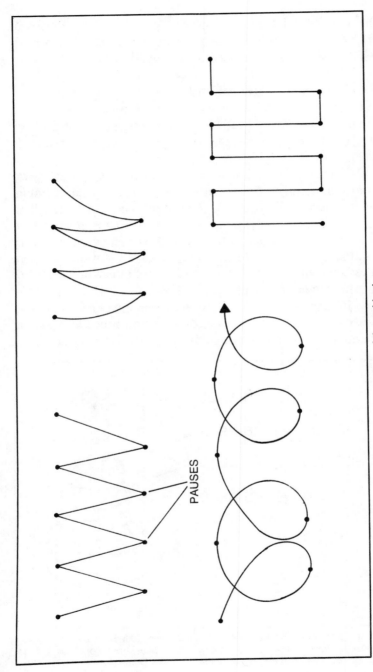

PAUSES

Fig. 8-10. Weave welding requires pauses to prevent undercutting of the weld edges.

a bit and chip off any slag at its end. Move forward ½ inch or so, strike an arc, and run the weave pattern back to your stop-off point. Then continue on with your welding.

At any time when maximum penetration is desired or needed, you will continue to use straight or stringer beads. Again, where distortion control is important, straight beads are also superior. The continued arcing in a small area during weave welding tends to build up heat more quickly than it does when a straight bead is run. When your practice with the various welds and beads begins to result in good looking work on a consistent basis, it's time for you to move on to fusion welds.

FUSION WELDING

Butt joints are generally ideal, though it never pays to put too much practice into a single type of weld joint at the expense of others. Eventually you will find a need to make a tee joint or a fillet weld on a corner joint.

Make your choice of materials for practice. I would recommend at least ¼-inch thick sheet metal, but you may have easier and cheaper sources of lighter or heavier materials. Stick with mild steel and, if you find holding an arc uncomfortably difficult, you can use an electrode designated as an E6013, or one listed as Airco does with their E7014 as an Easyarc.

Set up your two pieces of base metal, with their ends well supported on firebricks. Ground your welder, insert the electrode in its holder, and select the point at which you wish to begin welding. Drop your helmet into place and strike an arc, forming a tack weld first with your electrode. Depending on the length of your practice material, you may need to run a series of tack welds. These welds not only hold the base metal pieces in place for further welding, but also help to control distortion as heat buildup becomes great during running of the weld bead (Fig. 8-11).

While this tack weld may seem different from that made when gas welding, it is meant for the same purpose—holding the pieces in alignment. You can use an arc welder to make a traditional (no filler metal) tack weld by using a carbon arc

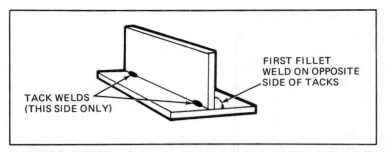

FIRST FILLET
WELD ON OPPOSITE
SIDE OF TACKS

TACK WELDS
(THIS SIDE ONLY)

Fig. 8-11. Tack welds can help keep distortion to a minimum. Tack, in fillet welds on T-joints, at both ends (and towards the center on longer welds). Then weld on the opposite side from the tack welds first. Tack welds are also useful as distortion controls on flat welds (courtesy of Hobart Brothers Company).

torch. The reason for making the tack welds with the electrode is that it gets to be tedious changing to a carbon arc torch to run five or six tack welds, and then swapping back to the electrode holder to run the actual finished weld.

You can run a stringer bead along the butt joint. When the weld has cooled and you've removed the slag, check the bead for characteristics. If it looks good, turn the practice pieces and make your second practice weld.

Your practice can go on until you begin to get really smooth stringer beads on each run. Practice pieces can be used and cut apart, though I would recommend that you make your cuts away from completed practice welds so that you can compare more recent welds with older ones.

Once the stringer beads are going smoothly, you will need two more pieces of metal, preferably a bit heavier. Widen your groove by beveling, U-jointing, or whatever process you prefer. If the metal is not much more than ¼ inch thick, a single V-joint will be best. Keep your included angle to about 90 degrees. Use a weave technique to make your weld after tacking the pieces together. Continue to practice until you develop skills with the various joints and weld beads that allow you to do well on most reasonable thicknesses of mild steels. Most of the skills you pick up for working on mild steel will transfer quite well to other metals, though you need some more practice in proper electrode selection and choice of amperage for other metals (such ranges are recommended by the manufacturers of electrodes).

A few tips from the experts can also help in maintaining good welds. When you change the size of the electrode you're using, make a couple of short bead runs on scrap stock in order to check out your amperage setting. Doing this before starting welding on good stock cuts down on waste. Make certain your practice includes vertical, horizontal, and over-head welding as well as flat welds. Practice both fillet and groove welds, too, along with multiple layering of welds, both weave and stringer bead, if you think you have to do much work in heavy materials.

VERTICAL WELDING

Vertical welding is the easiest of the so-called "hard" welding positions. It is usually started from the bottom of the weld and carried on up the weld, allowing the solidified part of the weld to hold the still molten material in place. Vertical welding may be done from either direction, but it is a lot harder when worked from the top down. For vertical welding, you use a slower welding speed and either an all-position electrode, or one that the manufacturer lists as good for vertical applications. Preference for welding in an upward direction is not simply because it is easier to carry out. Penetration is better. The upward direction is less likely to present the do-it-yourself welder, or even the professional welder, with a porous and slag-covered weld than is welding from the top down (Fig. 8-12).

The E6013 electrodes use a flux that helps the molten filler metal adhere to the base metal. It is not so sensitive to arc length that it will cause a welder many difficulties. Fast freeze electrodes can be used, but they tend to have great arc sensitivity. For down welding on the vertical, though, fast freeze electrodes are almost always best for most welders. Vertical welding recommendations include using as short an arc as is possible (consistent with arc freezing problems, which is why you'll find the E6013 type of electrode very good since it seldom sticks), keeping your electrode as close as you possibly can to 90 degrees in angle from the work surface, using the lowest amperage consistent with good penetration, and moving your electrode as rapidly as possible (while keep-

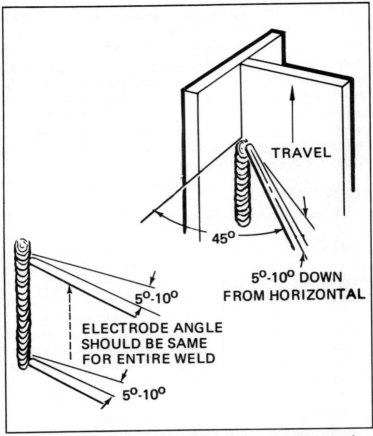

Fig. 8-12. Electrode angles for vertical welding need to be about 10 degrees from the horizontal, with the same angle maintained as closely as possible throughout the weld (courtesy of Hobart Brothers Company).

ing a wary eye on bead formation). These last two recommendations help to hold down the arc heat so you won't have to deal with a large molten puddle in the bead. Vertical weave welding follows along the same procedures and recommendations as does straight bead vertical welding.

OVERHEAD WELDING

Overhead welding is one process that is easier with arc welding equipment than it is with gas welding equipment, with enough practice on your part. Overhead welding, though, is still an unnatural welding position. Most people will probably want to avoid its use whenever possible. If you do find yourself needing to weld overhead, the primary techniques are similar to those used for vertical welding. There is a strong need for fast freeze electrode use whenever possible. The fast freeze electrodes give gravity less chance at dragging things down as you work because the puddle stays molten only for a very short time. Again, keep your amperage as low as possible and consistent with good penetration. If you use more amperage than is barely essential, you will have metal stalactitic shapes hanging from the weld. Keep the amperage down and make sure that you have good penetration of the base metal (Fig. 8-13).

HORIZONTAL WELDING

Gravity may cause you problems when welding along a horizontal line. Again, you must use the lowest amperage that

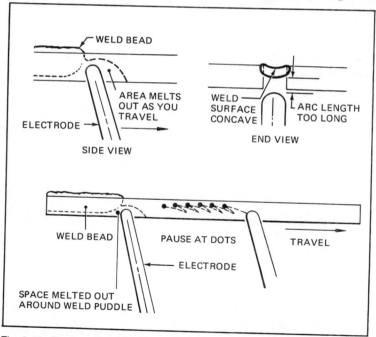

Fig. 8-13. Overhead welding requires great care to prevent loss of the puddle (courtesy of Hobart Brothers Company).

you can. Direct your arc slightly upwards and use a fast whipping motion, along with fast freeze electrodes, to help control placement of the molten material. If you use too much amperage, the bead will sag, becoming thicker at the bottom and looking very odd. The bead will be weakened by the sag (Fig. 8-14).

DIRECT CURRENT WELDING

Direct current welding is useful in situations where a more stable arc is needed. It also allows you to provide an arc with greater or lesser heat qualities.

Straight and *reverse polarities* regarding arc welding refer to the polarity of the electrode. On many machines you have to switch cables from one terminal to another, while on others you will find a polarity switch that allows instantaneous changeover. When you're using straight polarity, the electrode cable will be negative with the ground cable positive. Therefore, saying that the current is electrode negative is another way of saying straight polarity. Using straight polarity, you get about 65 to 70 percent of the heat given off on the positive side, or on the work area, with the electrode. Excellent weld penetration is one result of straight polarity dc arc

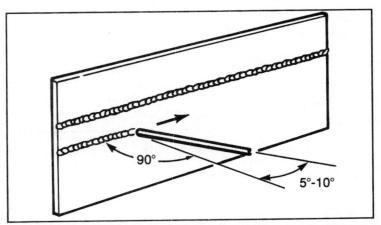

Fig. 8-14. Horizontal welds need an electrode held horizontal to the ground, and at about a 10-degree variance from a 90-degree angle to the work line. Weave welding techniques can be used to whip up and down and are often easier in thick material (courtesy of Hobart Brothers Company).

welding, and there is less chance of burning through the base metals (Fig. 8-15). For reverse polarity, your electrode will be positive and the workpiece negative, so the electrode retains more of the heat given off by the arc. Reverse polarity is handy when you need less penetration but more uniform transfer of the filler metal (Fig. 8-16).

ARC BLOW

Arc blow only becomes a problem when you're working with dc arc welding equipment. Current flow in the workpiece sets up magnetic fields which cause the arc to become unstable and make metal transfer difficult. Changing polarity doesn't provide any help, but sometimes simply moving the ground clamp to a different spot on the workpiece will break up the magnetic currents (which flow around the workpiece and tend to push the metal from the electrode away from the work surface) (Fig. 8-17). If it's possible, reducing the current will often help.

If moving the clamp and reducing the current don't provide a solution for arc blow, then try working the ground clamp cable around the workpiece. Take care to make sure the cable won't be too close to the actual weld and get overheated.

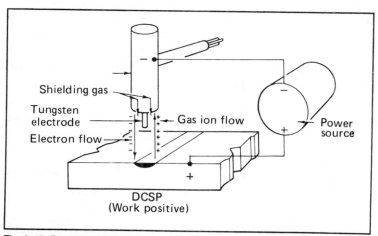

Fig. 8-15. Direct current, straight polarity connections (courtesy of Lincoln Electronic Company).

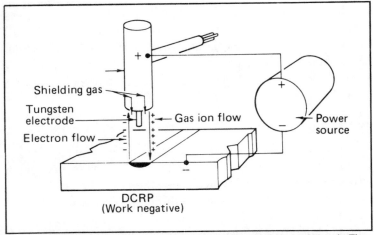

Fig. 8-16. Direct current, reverse polarity connections (courtesy of Lincoln Electric Company).

If you want to keep an arc blow problem from occurring there are two welding techniques you may use. First, tack weld the pieces in a few more places than you would normally bother with, and make the tack welds heavier. Use a very short arc and weld towards the tack welds.

A variation on this procedure involves tack welding the corners of the workpiece and then using a back-step welding technique to complete the weld. Start from a couple of inches in front of one tack welded corner and weld your way back to it. Move on a few more inches in front of that completed weld. Run your bead back to it. Continue back-stepping across the workpiece, and you should have no difficulty with arc blow (Fig. 8-18).

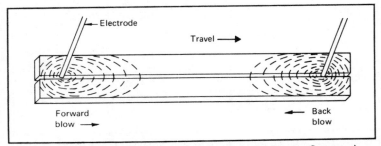

Fig. 8-17. Magnetic flux directions (courtesy of Lincoln Electric Company).

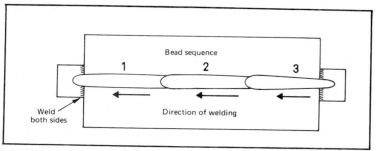

Fig. 8-18. Back-step welding (courtesy of Lincoln Electric Company).

ARC WELDING OTHER METALS

All your arc welding practice up to this point has been on mild steel. Occasionally, though, something that isn't mild steel needs to be welded—a cracked engine block of cast iron or a broken piece of wrought iron ornamentation. You may need to make a set of andirons or even a fire grate for the fireplace or woodstove from wrought iron or cast iron. If and when that time comes, you should know about the different techniques required for doing these jobs.

Stainless Steels

Arc welding of stainless steels is just as possible as gas welding of them. In most cases, though, it is better to simply braze weld or hard solder these steels. Internal changes in the metals can be exceptionally severe if you don't know the exact type and grade of stainless steel at the outset of the job. Even when the type is known, special heat treating equipment is often needed to counteract the internal changes in the metal. Unstabilized chromium-nickel stainless steel may undergo intergranular corrosion when subjected to heat only in the 800 to 1500 degree Fahrenheit range (427 to 816 degrees centigrade). If this is the stainless steel you are about to work on, make sure you hard solder it so that temperatures stay under those critical ranges.

Ferritic chromium or straight chromium stainless steels may lose their corrosion resistance when subjected to enough heat for fusion welding. They also have a tendency to become brittle. *Martensitic chromium* (also straight chromium) steels

179

also become brittle when subjected to fusion welding heat ranges. Both the straight chromium steels *can* be welded, even though your best bet is still braze welding or silver soldering.

Extra low carbon (ELC) content stainless steels are stabilized to prevent intergranular corrosion. These are nickel chromium types, so they do not become brittle quite as easily as the straight chromium types. If you are able to identify these grades, then special stabilized electrodes (such as Airco's 308 Easyfill) can be used to fusion weld them. Other stainless steel types, with columbium or titanium as stabilizers in the alloy, will require you to use an E347 electrode such as Airco's 347 (which is also designated for use on 308 ELC and 321 stainless). These electrodes are readily available. There is one valuable tip for identifying the stabilized stainless steels. You can almost bet that if the steel is used in a relatively high temperature application, it is one of the stabilized varieties. Whether you're working with milk cans or kitchen sinks, you should try to do the job of joining with braze welding or another lower temperature procedure.

Cast Iron

Cast iron is one of the hardest materials to arc weld. In fact, TIG welding or braze welding will almost always give you a better job. Sometimes, though, the strength associated with fusion welding is thought to be needed (though often braze welding will provide a stronger joint). Gray cast iron is the most common type found around the home and farm, and it is the one you're most likely to work on at some time. Arc welding produces an extremely rapid heat buildup and allows for exceptionally quick cooling once the arc is removed from the parts being welded. These quick temperature changes set up a lot of internal stresses in cast iron and cause carbon pickup in the weld from the base metal (cast iron contains much carbon that it gives up the filler metals during the various welding processes). Both the base metal and the weld become subject to easy cracking. Preheating is almost essential to a good fusion welding job, and some measures need to be taken to slow cooling.

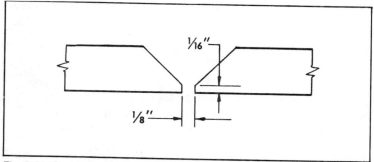

Fig. 8-19. Cast iron preparation (courtesy of Lincoln Electric Company).

First, when you have to fusion weld cast iron, make sure it is clean and free of rust and scale. Prepare a groove so that the root can fuse as easily as possible (Fig. 8-19). Preheat thoroughly. Preheating means you should obtain an oxy-MAPP or oxyacetylene torch to bring the overall preheat temperature up around 500 degrees Fahrenheit (260 degrees centigrade). Don't try to use a single gas propane torch or MAPP gas torch to heat a very large piece. The cast iron will dissipate the heat more quickly than the torch can produce it, long before the preheat temperature is reached. As soon as preheating is completed, begin your welding and continue until the weld is finished. The heat must be maintained.

Once the weld is completed, you need to keep the part from cooling quickly. If you've made a weld on the exterior of an engine block, you can cover the entire engine block with some heat-resistant material with a good insulating value. Cooling is thus retarded. You can place one of the foil space blankets over the engine, and then lay 3 to 6 inches of fiber glass insulation in roll or batting form over that. This procedure helps to slow cooling and prevent cracking.

You must use special electrodes for cast iron. Eutectic Corporation has their *Xuper 2240* for the job, though their engine block recommendation is *Xyron 2-24*. Since Eutectic calls the Xyron 2-24 a minimum heat electrode, it would also be my choice. Still, the Xuper 2240 is made for good penetration on contaminated surfaces while the Xyron 2-24 requires a nearly sterile work surface for best penetration. Much depends on the working conditions.

Aluminum

When you weld aluminum, the procedures involve the use of TIG equipment. For some small jobs, though, manual arc welding can be done. Aluminum presents two major problems to the arc welder. First, it forms oxides with a rapidity greater than almost any other metal. Second, it dissipates heat with just about the same speed that it forms oxides. The fast oxide formation can easily cause weld faults if the oxides manage to get included in the weld. The fast heat dispersal can aid oxide exclusion in the weld.

Oxides already present on the aluminum to be welded are removed with commercial cleaning solutions specifically prepared for welding uses. The rapid heat transfer problem can sometimes be solved by preheating with an oxyfuel gas torch, along with some form of insulation in and around the areas close to those that you need to weld.

Oxide formation when welding aluminum is further held down by using fluxed electrodes. Even thoroughly cleaned aluminum will be quick to form new oxides more rapidly than usual, as temperatures reach those of fusion welding. Welding temperatures for aluminum are not hot enough to burn off the formed oxides, as they are with mild steel.

OTHER ARC WELDER JOBS

Your arc welder is suitable for more jobs than simply welding together a few pieces of metal. Arc welders can do a good job of cutting metals with the proper accessories or using special electrodes meant for chamfering, gouging, cleaning, or cutting. Arc welders can be used for hard surfacing and are sometimes preferred over oxyfuel gas welding equipment for heavier hard surfacing jobs. Arc welders are fine for thawing frozen metal pipes if the correct techniques are used.

Arc Cutting

Electrodes used for chamfering and gouging can also be handy for straight cutting jobs with little or no change in equipment setup if you just follow the manufacturers' instructions. Nearly every type of coated electrode, except for those covered with iron powder, can be used for cutting ferrous metals. Simply holding the arc in a single place until the heat becomes intense enough to oxidize the metal. The reason iron powder electrodes do not cut well when used with an arc welder is that the coating used when cutting must serve as an insulator, allowing the electrode to move into the cut without shorting out.

Carbon arc cutting is similar to cutting with the electrodes. The arc is struck between the carbon electrodes (both electrodes are held in the torch), and the heat from that arc is used to sever the metal. The carbon arc torch will consume its carbon electrodes much more rapidly than when you use it as a welding or brazing torch. The cutting works best if you get a special cutting electrode holder and use it to hold the electrodes specially designed for use when cutting metals with an arc welder. Some of these carbon arc guns or torches can be bought with fittings for compressed air to blow away slag and molten metal as you cut. If you have access to an air compressor or a moderately large high pressure air bottle, such a fitting would be a reasonable addition to your arc welding and cutting shop. The special cutting and gouging electrodes are designed so that their arcs blow much of the molten material out of your way as you cut. A cleaner cut can be made using oxyacetylene or oxy-MAPP gas equipment for ferrous metals. Arc cutting works best in nonferrous material.

Hard Surfacing

Most hard surfacing is done to improve either impact or abrasion resistance, especially around homes and farms. For a couple of dollars worth of hard surfacing electrodes or rods, you can quadruple the life span of a shovel or an ax. Many formerly cheap tools are now fairly expensive and lend themselves well to hard surfacing in order to extend their life spans.

You may want to maintain a cutting edge without constant sharpening. At other times you may need to keep a single surface, with no sharp edge, from wearing too extensively. Another reason for hard surfacing is the protection or restoration of bearing surfaces.

Tools that need to have sharp edges like knives and axes require tool steel hard surfacing in most cases. Shovels, mattocks, and similar tools can be surfaced with other types of steels, as keeping exact dimensions is usually less important. Tool steels for hard surfacing come in both high temperature and moderate temperature applications. A moderate temperature alloy will probably do.

Shovels can be hard surfaced with an alloy containing *carbides* for greatest wear resistance. Both *tungsten* and *chromium carbides* are available. The tungsten carbides are suitable for use in extreme wear conditions where material coarseness won't cause a problem with the tool. Chromium carbides don't provide as much wear resistance for your tools as tungsten carbides, but they give you a smoother overall alloy deposit. They are far better for jobs where self-sharpening features are needed. The chromium carbides may be used on items like heavy-bladed knives. Impact resistance for both alloys is on the low side of the scale.

Hard surfacing is used to control wear from friction between two bearing surfaces. A type of bronze or some other alloy of copper will be bonded to the surface in order to decrease friction.

Arc Braze Welding

While it is generally accepted that brazing is more easily done with gas welding equipment, it is possible to do a fair job with arc welding equipment. The tool will be the carbon electrode holder. If you have no carbon electrode holder, you can use your carbon arc torch to provide heat for electric braze welding. While the carbon electrode holder is used to create an arc between the carbon electrode and the work surface metal, thus heating the base metal for the application of the braze welding rod, the twin carbon arc torch creates an arc between its own two carbon electrodes (ac only) and is harder to control. Carbon arc torch brazing shouldn't be attempted with very light metals. The intensity of the heat produced will virtually promise you a warped final result. While the carbon electrode holder also generates extremely intense heat, it is much more controllable. Your best bet, though, for brazing any light metals is to use gas welding or brazing equipment.

Arc braze welding is similar to gas braze welding, except that phosphor bronze and silicon bronze are the only commonly used filler metals with the carbon electrode holder. Joint design is identical. Applications of heat is the same as with any other heat source.

Thawing Water Pipes

Although the extensive use of plastic pipe for water runs on homes and farms has made it more difficult to thaw frozen pipes electrically, many pipe runs are still of metal like copper or cast iron. Sometimes these runs freeze in extremely cold weather.

Pipe thawing can be done with propane torches if care is taken to protect combustible surfaces. Arc welders may be better for thawing water pipes. The welders are excellent sources of low voltage current, allowing you to thaw even underground pipe lines with no digging. Heat application is relatively easy to control, but some dangers do exist.

When thawing pipe lines, you must create a circuit with a high resistance (that will be the ice-filled pipe). The circuit heats up, just as the high resistance wires do in your toaster. Timing and current requirements will vary with the size of the pipe and the material from which the pipe is made. Coppers heat less rapidly because of their greater conductivity than steel or cast iron, so more current is needed to thaw copper pipes.

You should use an *ammeter* to check line readouts when thawing pipe. Delivered amperage will be somewhat higher than the setting on the dial of your arc welder. At 200 amperes current (on the ammeter), copper pipe that is ½ inch in diameter will take about 3½ to 4½ minutes to thaw out. Copper pipe that is ¾ inch thick will take some eight to nine minutes to thaw out, while 1-inch pipe will take 14 to 18 minutes. Steel or cast iron pipe in large sizes will thaw with reasonable rapidity. Once the ferrous pipe goes over 1¼ inches in diameter, more than 200 amperes will be needed to do the job. Your 1¼-inch ferrous pipe will thaw in about 30 minutes at 200 amperes, but a 1½-inch pipe will require 240 amperes to thaw in the same length of time. Two-inch pipe will require a full hour to thaw when you apply 220 amperes.

No small welder should be used for pipe thawing if its rated duty cycle is under 30 percent. Care should be used on those lighter amperage models (duty cycles under 60 percent), with voltage outputs as low as 25. Specific operating instructions will be supplied by the machine's manufacturer. If your operating manual contains no such instructions for pipe thawing, and you expect to use the machine for this task, write to the manufacturer or the distributor.

Your pipe thawing job starts by making sure you've located the frozen section of pipe. You will probably be able to check the location by turning on the faucets in various areas until you've determined which are working.

Be sure you've got the welder turned off before you make any hookups with the cables. An arc from the cable connector to the pipe can be hot enough to destroy both the pipe and the connector. Clean the areas of the pipe where you will be attaching the connectors. Use coarse emery paper, steel wool, or a wire brush for the cleaning job.

Place your cable connections as close to the frozen pipe areas as you possibly can. Try to avoid empty stretches of pipe, as the heat generated in such areas could reach solder melting temperatures before the frozen section of pipe warms enough to thaw. If the pipe is inside a wall and empty, you might face the possibility of a fire hazard (Fig. 8-20).

Insulated pipe joints and plastic pipe sections will prevent a complete electrical circuit from being on the proper section of pipe. This could cause a short circuit condition because of the heavy current load, again creating a fire hazard. Check to make sure these types of pipe are not included in the possible circuit.

Leave all faucets open. Once the ice begins to melt, the first slight flow of the water will help to speed your thawing job. A full flow of water from the faucets will indicate to you that the thawing is completed.

Remember that the actual current draw will be higher than that shown on the indicator dial of your arc welder. This

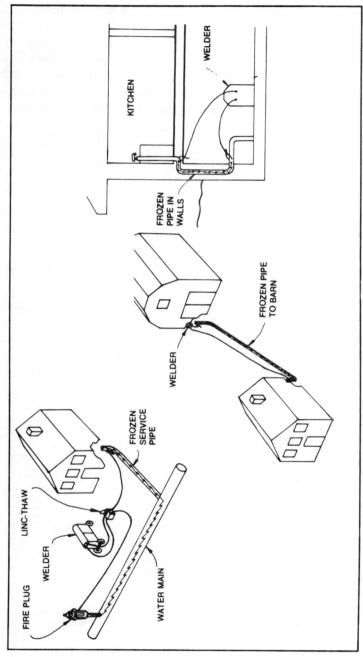

Fig. 8-20. Pipe thawing (courtesy of Lincoln Electric Company).

187

higher draw occurs because the circuit you are creating when thawing pipe is close to an actual short circuit across the welder terminals. If you operate the arc welder too long in this manner, it can sustain damage. It is imperative that you check with the manufacturer if pipe thawing instructions are not included in your manual. At no time should a constant voltage welder, with no control for limiting the current, be used for thawing pipe. You should never leave the welder unattended when pipe thawing is being done. Keeping an eye on things is a good way to prevent problems.

Chapter 9

Tungsten Inert Gas Welding

The welding of aluminum, magnesium, and stainless steel metals is considerably easier, and the resulting welds are much stronger, when you use *tungsten inert gas welding* (TIG). Basically, TIG welding is a form of electric arc welding that provides an oxygen or oxidefree weld by using a shield gas to prevent atmospheric oxygen from combining with the base metal and the filler metal (if a filler metal is used) (Fig. 9-1). The equipment you need to perform TIG welding chores will vary with the type of arc welder you have or intend to buy. If your arc welder operates only as an ac machine, then you will need to buy a high frequency stabilizer attachment as well as the regulators, hose, and TIG torch (Fig. 9-2). This high frequency stabilizer is essential in order to stabilize the ac arc; otherwise, it is impossible to hold an arc with the nonconsumable tungsten electrode.

If you already have an ac dc arc welder, then unless you want to use the ac setting for TIG welding, all you will really need is the torch/hose/regulator kit. You will need to add a cylinder of argon gas. Helium may also be used, but many of the TIG torch kits are supplied with regulators that will only work with argon. Argon is useful in more applications in most positions. Because of argon's weight, though, helium is better when overhead welding must be done.

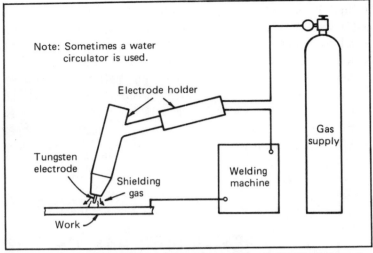

Note: Sometimes a water
circulator is used.

Electrode holder

Gas
supply

Tungsten
electrode

Shielding
gas

Welding
machine

Work

Fig. 9-1. The basic TIG welding outfit (courtesy of Lincoln Electric Company).

HIGH FREQUENCY STABILIZER ATTACHMENT

The TIG high frequency stabilizer attachment also makes some other types of ac welding easier, especially if you are using low hydrogen or other electrodes which make it difficult to maintain an arc. The high frequency attachment makes it possible for you to start an arc without touching the workpiece and then stabilize the resulting arc. Your biggest problem when using the high frequency attachment is that there is a major reduction in welder duty cycle when it is in use. Sears' high frequency attachment has a 30 percent duty cycle. It also tends to reduce the duty cycle of the welder to which it is attached if that welder isn't a heavy-duty model rated at 60 percent or more.

The reduced duty cycle isn't much of a problem in TIG welding. Welding speed can be increased because there is no slag in the weld to cause problems.

DCSP AND DCRP

The basic TIG welding process involves the arc being struck from the nonconsumable tungsten electrode, which is held in a special holder or torch. The inert gas is fed to the arc through the electrode holder and forms a shielding pocket

190

around the arc, keeping out air. Filler rod or filler metal can be fed in with the other hand when and if it is needed. One of the great advantages of TIG welding is the possibility of using strips of the base metal as filler material. Since the alloy has no need to be coated to cut down on oxidation, the actual base metal can be matched exactly. Great strength and a near exact color match result. Working with filler metal also reduces the need to test metals in order to check alloy formation This need is still present when the metals to be welded are temperature sensitive.

When you begin to install your high frequency arc stabilizing equipment, even assuming correct installation of your basic arc welder, you will find a few requirements not normally associated with welding equipment. All power and communication lines within a radius of 50 feet must be enclosed in metal conduit to reduce interference with television

Fig. 9-2. The ac arc stabilizer high frequency oscillator (courtesy of Lincoln Electric Company).

and radio reception. A double ground of the conduit is recommended, with one at the welding machine and one about 50 feet away. Use electrical jumpers at any point where you will find a joint in the conduit. If radio or television interference later becomes a problem, you will find it necessary to reset the points in the high frequency. These points will need to be reset after every 150 to 200 hours of operation. Each unit will need a slightly different gap which will be specified by the attachment's manufacturer. Make sure your attachment is unplugged before you make any point adjustments.

Most TIG welding will be done with either high frequency ac or direct current, straight polarity (DCSP). DCSP simply means your welder is set so that the workpiece is positive, and the electrode is negative. The electron flow is from the electrode to the work, presenting the greatest amount of heat on the work surface while leaving the electrode fairly cool. If direct current, reverse polarity (DCRP) is used, the electrode holds much of the heat (about 65 percent). The resulting welds are necessarily shallow, so that the thickness of the metals that can be welded with DCRP is limited, in aluminum, to about 0.05 inches.

With heavy-duty TIG welding machines, the welding heat becomes so intense that a water spray is needed to cool things down. With the equipment most do-it-yourself welders will be using, the lowered duty cycle and air cooling will do a sufficient job of keeping the heat rise under control.

STRIKER PLATE AND ARGON FLOW

With your arc welder set up and the TIG attachment in place, clamp a tungsten electrode in the electrode holder. Joint preparation is even more important for TIG welding than with other types. A close match at the joint, while leaving sufficient room for the bead root, is essential to a strong job. Clean surfaces are also of great importance. When your workpieces are in place, if the material being welded is aluminum or magnesium, you will need to select a chunk of copper as a *striker plate*. The striker plate or block is placed against the starting portion of the weld joint and used as an arc striker, so that contamination from the atmosphere doesn't occur even at the start of the weld. Using the striker plate as a precaution against weld contamination isn't essential with metals which oxidize less rapidly than aluminum and magnesium.

Once the striker plate is set, you're ready to start the flow of argon. Regulators for such shield gases come in pairs, just as do those for fuel gases and oxygen, but there is a major difference in the gauge on the working regulator (Fig. 9-3). The working regulator gauge does not show pounds per square inch, but instead gives you the flow in cubic feet per hour. The cylinder regulator gauge is standard and shows the pounds per square inch remaining in your cylinder (argon is

Fig. 9-3. Argon is used at cubic feet per hour with a flowmeter rather than with a pounds per square inch regulator (courtesy of Airco Welding Products, Murray Hill, NJ).

usually pressurized to 2300 psi). Argon flow to the arc depends on the thickness of your weld, joint design, filler rod diameter (if filler rod is used), or the thickness of the filler strips cut from the parent metal. Mild steel that is ⅛ inch thick, with a square groove joint, requires a flow of 12 cubic feet per hour. Aluminum of the same thickness and joint design requires more shielding gas. Figures in Tables 9-1 through 9-3 are only starting points for determining shielding gas flow. The actual required gas flow will depend on your welding skill and speed. You must also think about cross currents of air and sudden drafts in the area of the gas shield arc.

STRIKING THE TIG ARC

There are a couple of ways to strike a TIG arc. The method used depends on whether or not a high frequency arc stabilizer is used for ac and dc work. If the high frequency unit is attached and operating, you will need only to move the tungsten electrode tip down close to the work surface. The arc will strike itself before actual contact is made with the metal. Hold the arc at the starting point until you get a clean, bright puddle of molten metal about twice the width of the electrode's diameter. For a high frequency stabilized arc, you do not need a striker plate, no matter what metal is being welded. If no high frequency arc stabilizer is to be used (dc only), the arc will not strike until the electrode tip actually

Table 9-1. Amperage, Electrode, and Gas Flow
Charts for Brass Alloys (courtesy of Sears, Roebuck and Company).

Metal-Brass Alloys						
metal thickness	type of weld joint	tungsten electrode diameter	ac-hf welding current (amps)	dc-sp welding current (amps)	shielding gas-argon cu. ft./hr.	filler rod diameter
1 /16″ (.062″)	Square Groove	1 /16″ (.062″)	105-155	85-125	15	1 /16″ (.062″)
1 /16″ (.062″)	Fillet	1 /16″ (.062″)	105-155	85-125	15	1 /16″ (.062)
1 /8″ (.125″)	Square Groove	1 /16″ (.062″)	145-190	115-150	15	3 /32″ (.093″)
1 /8″ (.125″)	Fillet	1 /16″ (.062″)	145-190	115-150	15	3 /32″ (.093″)
3 /16″ (.187″)	Square Groove	3 /32″ (.093″)	180-200	145-195	20	3 /32″ (.093″)
3 /16″ (.187″)	Fillet	3 /32″ (.093″)	180-200	145-195	20	3/32″ (.093″)
1 /4″ (.250″)	Square Groove	3 /32″ (.093″)	—	160-200	25	1 /8″ (.125″)
1 /4″ (.250″)	Fillet	3 /32″ (.093″)	—	160-200	25	1 /8″ (.125″)

AC-HF (ALTERNATING CURRENT-HIGH FREQUENCY STABILIZATION)
DC-SP (DIRECT CURRENT-STRAIGHT POLARITY)

Table 9-2. Amperage, Electrode, and Gas Flow Chart for Mild Steel (courtesy of Sears, Roebuck and Company).

Metal-Mild Steel						
metal thickness	type of weld joint	tungsten electrode diameter	ac-hf welding current (amps)	dc-sp welding current (amps)	shielding gas-argon cu. ft. /hr.	filler rod diameter
1 /32" (.031")	Square Groove	1 /16" (.062")	95-125	75-100	10	1 /16" (.062")
1 /32" (.031")	Fillet	1 /16" (.062")	95-125	75-100	10	1 /16" (.062")
3 /64" (.046")	Square Groove	1 /16" (.062")	115-150	90-120	10	1 /16" (.062")
3 /64" (.046")	Fillet	1 /16" (.062")	115-150	90-120	10	1 /16" (.062")
1 /16" (.062")	Square Groove	1 /16" (.062")	120-170	95-135	10	1 /16" (.062")
1 /16" (.062")	Fillet	1 /16" (.062")	120-170	95-135	10	1 /16" (.062")
3 /32" (.093")	Square Groove	3 /32" (.093")	170-200	135-175	10	3 /32" (.093")
3 /32" (.093")	Fillet	3 /32" (.093")	170-200	135-175	10	3 /32" (.093")
1 /8" (.125")	Square Groove	3 /32" (.093")	180-200	145-200	12	1 /8" (.125")
1 /8 " (.125")	Fillet	3 /32" (.093")	180-200	145-200	12	1 /8" (.125")

ac-HP (ALTERNATING CURRENT-HIGH FREQUENCY STABILIZATION)
dc-SP (DIRECT CURRENT-STRAIGHT POLARITY)

touches the work surface. For most of your work, you will find a tungsten electrode the easiest with which to start an arc. Arc length in TIG welding, as a rule of thumb, should be about 1½ times the diameter of your tungsten electrode. This arc length and puddle width should give you a good, narrow bead and deep penetration.

Your electrode holder is angled so that you can get a good view of the work. Some experts recommend a 90-degree angle for butt joints, but I prefer at least a 15-degree slant from the perpendicular so that I can better view both the arc and the work. For tee joints, try a 45-degree angle from each piece of the work. Slant the electrode about 10 degrees in the direction of its travel (Fig. 9-4). None of these angles is an

Table 9-3. Amperage, Electrode, and Gas Flow Chart for Aluminum Alloys (courtesy of Sears, Roebuck and Company).

Metal-Aluminum Alloys						
metal thickness	type of weld joint	tungsten electrode diameter	ac-hf welding current (amps)	dc-sp welding current (amps)	shielding gas-argon cu. ft. /hr.	filler rod diameter
3 /64" (.046")	Square Groove	1 /16" (.062")	40-60	—	20	1 /16" (.062)"
1 /16" (.062")	Square Groove	3 /32" (.093")	70-90	—	20	3 /32" (.093")
1 /16" (.062")	Fillet	3 /32" (.093")	70-90	—	15	3 /32" (.093")
3 /32" (.093")	Square Groove	3 /32" (.093")	90-115	—	20	3 /32" (.093")
3 /32" (.093")	Fillet	3 /32" (.093")	90-115	—	15	3 /32" (.093")
1 /8" (.125")	Square Groove	3 /32" (.093")	115-140	—	20	1 /8" (.125")
1 /8" (.125")	Fillet	3 /32" (.093")	115-140	—	20	1 /8" (.125")
3 /16" (.187")	Vee Groove	1 /8" (.125")	190-200	—	20	1 /8" (.125")

ac-HF (ALTERNATING CURRENT-HIGH FREQUENCY STABILIZATION)
dc-SP (DIRECT CURRENT-STRAIGHT POLARITY)

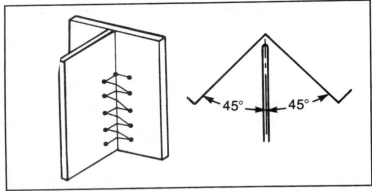

Fig. 9-4. Electrode position and weld weave for tee joints (courtesy of Hobart Brothers Company).

absolute when TIG welding. A variation of 5, 10, or more degrees is acceptable if you can work easier (Fig. 9-5).

TIG arc welding differs from standard arc welding in the welding *breaking* technique you will need when approaching the end of a bead on aluminum. Increasing the speed of your welding just before breaking the arc will decrease the chances of the crater cracking. Restriking the arc and coming back to fill the crater with molten metal will also help.

TUNGSTEN ELECTRODES

Tungsten electrode selection is somewhat simpler than the selection of standard electrodes because of the smaller number of types. The type of tungsten electrode you select to produce your nonconsumable arc depends in large part on the type of current you use to carry out the TIG welding. There are three types of tungsten electrodes: *pure tungsten, thoriated tungsten* (which has a tiny percentage of thorium alloyed into the electrode), and *zirconium tungsten* (a small percentage of zirconium is alloyed into the electrode). The pure tungsten electrodes may be used with any TIG welder and are lower in price than the alloyed electrodes. These electrodes are excellent when you are welding aluminum. Pure tungsten electrodes are best when used with alternating current. The tip color code for pure tungsten electrodes is green. Thoriated tungsten electrodes come with a double color code; electrodes with 1 percent thorium have yellow tips, while the ones with 2 percent thorium have red tips. Thoriated tungsten electrodes are useful at lower amperages and leave less contamination in the weld than pure tungsten electrodes if they happen to get shorted out. Zirconium tungsten electrodes are color coded brown at their tips. They also contaminate welds when shorted out less than pure tungsten types. The zirconium tungsten electrodes are, like pure tungsten, best used with ac welders at higher amperages.

Tungsten electrodes, no matter what their alloy is, come in two finishes. Standard finished electrodes are etched, or chemically cleaned, and sized to a reasonably close diameter for use with TIG electrode holders. Ground tungsten electrodes cost about 50 percent more than the etched electrodes. They are of little interest to the do-it-yourself welder because they are meant for extremely precise work. Electrode diameters range up to ¼ inch. The lengths vary from 3 inches to 2 feet.

FILLER METALS

The best term to use is filler metals; filler rods are not essential to TIG welding. While basic arc welding uses a

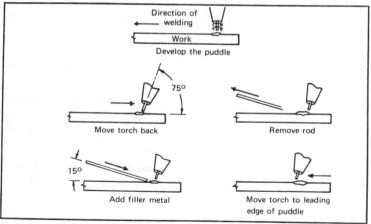

Fig. 9-5. Manually feeding filler metal into the weld puddle when TIG welding (courtesy of Lincoln Electric Company).

consumable electrode that must be included in the welding process in order to fuse the base metals, TIG welding offers an advantage no other readily available welding process can offer. You have a chance to use strips cut from the base metal as filler metal in a weld. This use completely removes any difficulty in determining the correct filler alloy to use, and always allows you to make sure that you've got the right filler alloy for any base metal alloy. Decorative welding applications are greatly enhanced by the possibility of very close color matches.

TIG WELDING OF METALS

Your TIG welding skills should first be tried on aluminum. These skills start from the point where basic arc welding skills leave off. You can move on to metals that are even harder to weld. For metals other than aluminum and magnesium, you can simply use straight polarity dc.

Magnesium

When you're getting ready to weld magnesium, make sure you have the special fire extinguisher on hand. Helium is generally somewhat preferred by professionals when welding magnesium. It produces a somewhat hotter arc than argon does, but most do-it-yourself welders will probably have regulators and flowmeters that work well only with argon. Penetration is slightly reduced when you use argon, but otherwise there is no major problem with it.

A 2 percent thoriated tungsten electrode or a zirconium electrode will give you less weld contamination and greater current carrying capacity than a pure tungsten electrode will. Make sure the magnesium surfaces to be joined are as clean as possible. Make sure the filler rod doesn't carry any contamination into the weld.

You can weld magnesium by using the same procedures you used for aluminum. Use a striking plate to start the arc. Speed up the welding pace as the end of the bead nears.

Stainless Steels

Welding stainless steel with TIG equipment is easier than welding aluminum because ferrous metals don't dissipate the heat as quickly. The stainless steel will change color to let the welder know how close it is to the puddling point, while aluminum just collapses without changing color. If you find your puddle ripples disappearing and the metal starting to turn a purplish color, then you're using too much heat. Keeping a close watch for any change to a dark purple will result in a much better weld in stainless steel. Use straight polarity dc for a stable arc because you don't need oxide cleaning capabilities here. Argon is used as the shield gas. Use a high weld travel speed (about 10 inches per minute in ⅛-inch thick stainless steel). This extra speed helps keep the heat buildup in the base metal as low as possible. Whenever possible, the 2 percent thoriated tungsten electrode should be used.

Carbon and Low Alloy Steels

Carbon and low alloy steels are classified as metals that are hard to weld with standard arc or gas welding equipment. The use of straight polarity dc TIG welding will give you a good weld. Carbon and low alloy steels are easy to weld when compared with aluminum and magnesium. The changing colors of the ferrous metals give you an indication of when the puddle is about to form. Additionally, there is no real need for strict joint preparation to protect against oxide formations in the weld. Most oxides will burn off as the metal puddles.

Using a high frequency attachment to stabilize your arc, even with dc, is a help as the electrodes used for low alloy steel welding need to have sharp points. The sharp point is ruined if it comes in contact with the work because the metal might cling to it. The electrode will then cause an unstable arc. Being able to strike an arc without touching the work surface is handy. Use 2 percent thoriated tungsten electrodes. Sharpen the electrodes at the start of the weld and every so often during the welding process. Use TIG welding rods or a strip of base metal for filler material. Oxyacetylene mild steel rods cause spattering.

Copper and Copper Alloys

Welding copper and copper alloys requires very little oxide cleaning action from the arc, so you are able to use the stable arc of straight polarity dc. Helium shielding is preferred, but again most home and farm welders have equipment that will only handle argon. Use 1 percent thoriated tungsten electrodes for most copper welding, though silicon bronze is best welded with a 2 percent thoriated tungsten electrode. For silicon bronze, only use the 1 percent electrode in a pinch. Keep a fast arc travel speed to prevent a condition known as hot shortness. The metal becomes brittle as the temperature reaches red heat, so you must keep this red heat area as small as possible. Because of this brittleness at red heat, copper alloys also require good support during fusion welding. Use a small weld puddle and a 60-degree included angle between the electrode and filler metal rod.

If the expense of the shielding gas doesn't become too much, TIG welding equipment can be used to do nearly every welding job you encounter. If you have a metal that is hard to weld with ordinary arc welding equipment, and you have access to TIG welding equipment, use the TIG welding gear. When standard arc welding suffices for nearly all the work you are doing, though, there is no real point in spending the extra money for TIG equipment.

Chapter 10
Soldering

Almost all of the available soldering techniques are exceptionally handy ways of joining metals for many purposes. Most require little equipment and are relatively simple to do.

The field of soldering can be divided into *silver soldering, torch soft soldering*, and *electronic soldering.* Silver soldering (or silver brazing) requires you to use a torch and depends on capillary action for the flow of filler metal. It provides the strongest joint of all the soldering methods and requires the most heat. Torch soft soldering uses essentially the same heat sources as silver soldering, but it does not give as great joint strength because the filler metal used is softer and weaker. Torch soft soldering also depends on capillary action for the flow of the filler metal, though the metal compounds used are quite different and not nearly as strong as those used with silver soldering. Much less heat is also required, as the softer metal alloys melt at much lower temperatures. Electronic soldering is usually done with a lightweight pencil style soldering iron or a soldering gun. It is primarily used to make secure joints capable of good electrical current flow. Electronic soldering alloys are similar to those used for torch soft soldering, though not quite identical.

SILVER SOLDERING

Silver or hard soldering uses temperatures in excess of 800 degrees Fahrenheit (427 degrees centigrade) at the joint before the nonferrous filler metal is added. Joint clearances are critical. They must not impede capillary action, which is the only way the filler metal flows. Capillary action or flow is the result of adhesion between molecules, causing a liquid to flow along a solid surface. Capillary action will allow the liquid filler metal to flow into any properly designed joint (usually tubular or cylindrical in shape, but far from always). Heat seals are thus possible in pipe runs and other places. If pipe in a run is to be silver or hard soldered, you must first consider the melting temperature of the base material used in the pipe. Silver soldering (also known as silver brazing) is seldom, if ever, used and the extra expense of the silver compounds just isn't justified.

Silver solder is useful in spots where soft solder cannot provide a strong enough joint, and where joint design can't meet the criteria for braze welding. Silver solder can be used to join steel, bronze, aluminum, magnesium, model, cast iron, and other dissimilar materials with relative ease, while still providing a moderately strong joint. Stainless steels, where joint strength of no more than 15,000 psi is required, can easily be joined with silver solders such as Eutectic Corporation's 157 or 157B (the B materials are for use with a soldering iron). The melting point for EutecRod 157 is only 425 degrees Fahrenheit (218.5 degrees centigrade), making it a soft solder. For true silver brazing or hard soldering, you would use a material such as Xuper 1020 XFC with a melting temperature of 1050 degrees Fahrenheit (566 degrees centigrade).

Even though the cost of silver alloys is high per unit of weight, the amount needed is so small in most applications that the overall speed of operation, and lower heat needs, make it reasonably economical for most jobs. Generally, an oxyfuel gas torch is considered to be necessary, but today many of the newer MAPP gas torches will provide sufficient heat for use on small to moderate sized parts being silver brazed or hard soldered (MAPP gas burns in air at about 3600 degrees Fahrenheit (Fig. 10-1).

Melting Points

Hard soldering alloy melting points will range from 1000

Fig. 10-1. Brazing a scrolled shelf to lightweight metal (courtesy of Airco Welding Products, Murray Hill, NJ).

to 1500 degrees Fahrenheit (538 to 816 degrees centigrade). The base metal must be hotter than the alloy's melting point before the filler metal is added to the joint. Otherwise, capillary action cannot occur. The lower temperature alloys will be used with metals like aluminum. The manufacturer will specify a liquid temperature for each alloy it makes.

Cleaning

Joints for silver soldering must be carefully cleaned, usually requiring both mechanical and chemical cleanings.

Use a fine file, emery cloth, or a fine grade of steel for the mechanical cleaning. Keep on cleaning until all surfaces are bright and shiny. Coat the surfaces to be joined with a flux of the correct composition for the metals being joined and the alloy used to make the joint. A tight fit is needed at the joint, but too tight a fit will impede capillary action. Keep joints in copper to 0.003 to 0.006 inches of clearance (double that figure for joints in aluminum).

All dirt, grease, and other grime should have been removed during mechanical cleaning. The chemical cleaner or flux is designed to remove oxides and prevent or reduce their reformation during the soldering process. After the flux is added, heat is applied to the joint. If you've chosen the correct flux, it will start to bubble and become fluid (depending on the type of flux you use, you may find it looking dried out just before it reaches the fluid stage rather than bubbling) just as the base materials reach soldering temperature. Flux is usually a paste which you must apply with a small brush. Once the flux becomes fluid, you add the filler metal to the joint. If your joint has been properly cleaned, fluxed, and heated, the filler metal will spread rapidly over the entire fluxed area. As soon as the joint is completely covered with filler metal, remove the heat from the joint. Excessive heat will create joint porosity. It isn't because the extra heat will weaken the bond between the metals and cause early joint failure.

Once joint cooling is complete, scrub the excess flux off with hot, soapy water. Use a file to remove any extra filler metal that might have built up. If you don't remove the excess flux, it could corrode the metal around the joint, depending on the metals joined and the type of flux. Be sure to get all the flux off the finished work. Even if the flux isn't the corrosive type that will cause corrosion problems, it generally is quite ugly. Almost any types will cause any paint laid over it to peel quickly.

Flux specifications for the type of filler alloy and the kinds of metals to be joined can be supplied by your welding distributor. Filler metals come in a variety of shapes, so you're not limited to rods or electrodes. These filler alloys are generally available as rings (to fit in a pipe joint), rods,

204

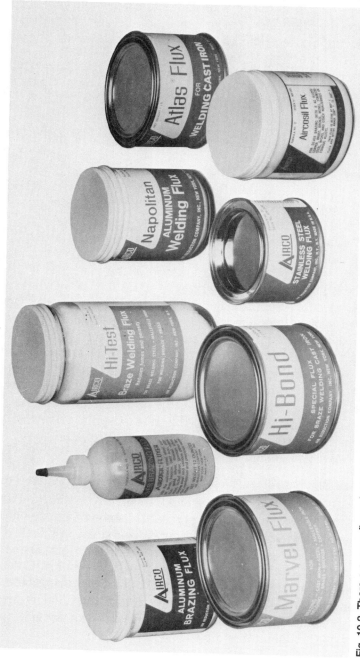

Fig. 10-2. There are many fluxes for brazing, braze welding, and welding (courtesy of Airco Welding Products, Murray Hill, NJ).

wires, and sheets. The wide range of shapes allows you to shape the filler alloy almost exactly to the joint shape, thus reducing brazing time and the possible need for excessive application of heat to melt enough filler metal. For special jobs, you can even get silver soldering alloys in powder form (Fig. 10-2).

Joint Design

Joint designs for silver soldering are not at all like those used with fusion and braze welding. In those instances, the bond you get between filler metal and base metal creates a very tough joint with little or no mechanical support and no real need for overlap at the joint. With silver soldering, as with soft soldering, the joint design must be completely different because the filler alloys provide little strength. Essentially, you'll find a use for two types of silver brazing joints, with variations for the shape of the materials being joined. Butt joints, again with only 0.003 to 0.006 inches of clearance for filler material, can be used. They often are used in an applications like joining the ends of bandsaw blades (you'll find it best to grind at an angle across the blade so that there is more butt surface for the filler metal to grip) (Fig. 10-3). The preferred joint for use with silver brazing is the *lap joint*. A lap joint provides a much larger area for the filler metal to bond to, and it is applicable for pipe joints (these are considered tubular lap joints) (Fig. 10-4). Capillary action will fill in the entire lap if cleaning is thorough, flux is spread well, and your gap is kept to recommendations.

Silver Brazing Alloys

Silver brazing alloys can vary in silver content from as little as 10 percent up to 80 percent. In some cases the only other metals in the alloy will be copper and zinc—in a range of percentages to give you the qualities you need to join particular metals. In other cases, cadmium is part of the alloy. Other elements sometimes added are nickel, tin, and manganese. Manganese that is alloyed in tends to take the temperature range up into braze welding areas, while alloying tin tends to drop the temperatures a bit.

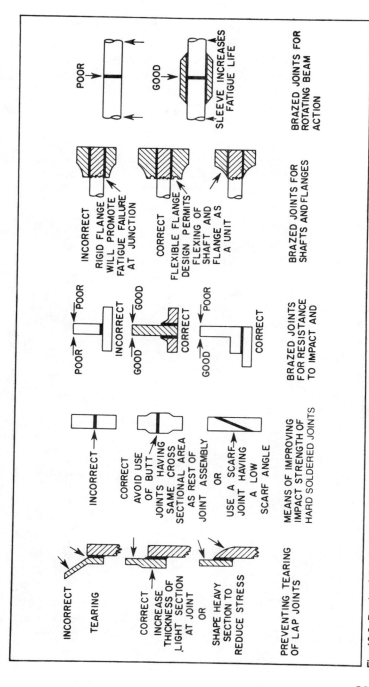

Fig. 10-3. Brazing joint designs (courtesy of Airco Welding Products, Murray Hill, NJ).

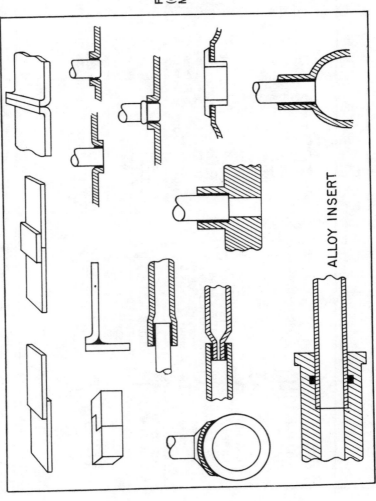

ALLOY INSERT

Fig. 10-4. Pipe joints at lap joints (courtesy of Airco Welding Products, Murray Hill, NJ).

208

The solder manufacturer will recommend the particular alloy for a particular type of metal to be joined, so that it is easily possible for the hard soldering beginner to do things right. With alloy information on hand, all you need is the correct flux for the materials being worked. Then you can, with care be assured of a good silver soldering job.

Torch Flames

When you are using oxyfuel gas torches, some filler alloys require a carburizing flame, using from 2X to 3X (Fig. 10-5). Others need a neutral flame (Fig. 10-6). Generally, when you find that a manufacturer's instructions don't require a carburizing or reducing flame, you'll be perfectly safe to use a neutral flame.

For brazing with MAPP gas torches, you have no neutral/carburizing adjustment. The torch flame is basically neutral and can usually be used with almost any type of filler alloy. These small single gas torches, even with modern torch head designs that greatly improve heat production, are not suitable for large silver brazing jobs. The amount of heat produced just isn't enough to heat large sections of stainless steel pipe, and it won't work well with aluminum of even moderate size. Rapid heat dissipation becomes a problem with aluminum. Tubular pipe sections have a tendency to draw cooling air through their centers, making it difficult to maintain heat levels needed. When silver brazing, the ends of

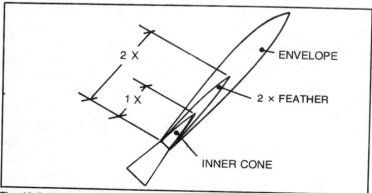

Fig. 10-5. A 2X carburizing flame.

Fig. 10-6. Neutral flame.

the pipe should be plugged to cut down on air flow if possible (Fig. 10-7).

When heating a joint to be silver soldered, play the torch flame over the base metal in a manner that heats the entire joint. Do not play the flame on the filler metal. The alloy will then melt long before the base metal is hot enough to bring about capillary action and give you the flow of filler material into the joint that you need for good bonding (Fig. 10-8).

Problems

The only way to tell when the heat range is correct is for you to choose the flux and then keep an eye on it, so you can tell when the appropriate temperature is reached. Too little heat is another major problem with hard soldering work. If you apply filler metal before the base metal is up to the proper temperature, the filler metal will ball up on the surface of the base metal. It will simply ball instead of flowing into the joint. This low heat condition is very easy to recognize and quite easy to correct, though the solution is not the obvious one of immediately applying more heat. First, you must let the joint cool down well below the melting temperature of the flux. Next, clean off the balled filler metal and reflux the joint. Apply heat as you normally would. The joint can be made with

Fig. 10-7. Pipe brazing (courtesy of Wingaersheek Inc.).

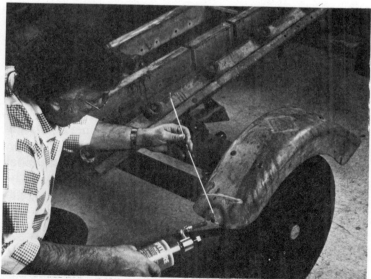

Fig. 10-8. The flame from the torch is applied to the joint and not to the filler metal (courtesy of Wingaersheek Inc.).

211

no more difficulty, as long as your second heat application warms things up enough. Most people with little experience in hard soldering, after insufficiently heating a joint, tend to apply too much heat the second time and end up with an overheated, porous joint. Keep your eyes on the flux for a good job every time (Fig. 10-9).

If the balls of filler metal from an underheated joint adhere to the base metal, you'll usually be able to just wipe them off with a rag. Occasionally you'll need to use a wire brush before refluxing.

SOFT SOLDERING

Soft soldering cannot provide you with a joint of great strength, as welding and braze welding do, but it can seal pipe joints and make very good electrical joints. Soft solder cannot be used to support any material where stresses are not light. A soft solder joint that is mechanically supported can provide a water seal in pipes or an electrical current passing joint in radios or other electronic equipment.

The basic soft solder joints I cover in this section will allow you to sweat solder pipes for plumbing and to solder electrical connections. Use the appropriate equipment in each case to make the job as easy as possible and to prevent damage to sensitive electronic components.

Very little residential and farm wiring requires the use of soldered splices or terminals, so there is little need to go deeply into that work. The tools and processes you would use are similar to those used with electronic equipment, only larger. If you need to solder an electrical connection larger than one suitable for use with a number 22 wire, simply transfer the sizes of the equipment needed and go ahead.

The primary technique for soft soldering copper pipe joints is known as sweat soldering. Essentially, you'll find this technique the same as that used to get silver solder into joints. A cleaned, fluxed joint is made and heated, and then the filler material is applied. The filler material then flows into the joint by capillary action, and the joint is sealed. When soft solder is used for a pipe joint, the joint strength is quite low. Resistance to general in-home vibration and light tugs and

212

Fig. 10-9. When the flux reaches the correct temperature, it will bubble.

twists is good because the filler metal is soft and allows the joint to give a bit under light stress. Sealing is excellent.

For sweat soldering, use a filler material of nonferrous metal with a melting temperature of less than 800 degrees Fahrenheit (427 degrees centigrade). The base metals, as in silver soldering, are not melted. Your joint strength depends on bonding of kinds other than fusion on the base metals. Soft soldering produces the weakest joints of all the metal joining processes covered in this book (those that use heat). The joints of greatest strength are produced by fusion welds. If soldered joints are to be subjected to stresses for which the soft solder hasn't enough strength, you'll need to use special joint designs or other means of mechanical joint support to prevent joint separation.

Soldering Temperatures

Soft soldering's low temperatures mean that you can make joints in materials that might otherwise be damaged by heat, or that you can make joints in metals next to materials that would be damaged by the heat flow from any other joint making process. Generally, no soft soldering will be done at temperatures of more than 700 degrees Fahrenheit (371 degrees centigrade). The scum (which usually forms on the surface of the soldered joint) becomes soluble in the solder at that temperature and will cause a poor seal, easy cracking, or other joint problems. Many soft solders melt at temperatures as low as 425 degrees Fahrenheit (218.5 degrees centigrade). You might locate some special purpose soft solders which melt at a point not far above that at which water boils.

The compounding of soft solder, consist mainly of varying percentages of lead and tin and affects the range over which the solder is plastic. A solder with a short plastic range might stay solid until its temperature reaches 500 degrees Fahrenheit (260 degrees centigrade). Then it will turn completely liquid at 515 degrees Fahrenheit (268.5 degrees centigrade). Quick melting solders with short plastic range tend to "freeze" on the joint rapidly because solder starts to return to its solid state when the temperature drops back to its melting point. It is solid when the temperature drops down to what is known as the *solidus point.* Metals have two temperatures involving melting points. The solidus temperature is the highest temperature at which the metal remains solid; *liquidus* is the first temperature at which the metal will actually begin to flow as a liquid. Once the temperature drops back down under the solidus point, the solder joint can be moved. Before the solder reaches that point, the joint has no strength at all. Any movement will result in a poor joint (this is called a fractured solder joint in electronic soldering and is often a major problem for those just learning how to solder complex equipment).

Between the solidus and liquidus is the plastic range of the solder. Some solders used for one job require a short plastic range, while others must have a much wider plastic range. The solders used for automobile body leading require a very wide plastic range so that they can easily be paddled over the surface of the damaged body area with no reheating required. Reheating and extra paddling of solders may coarsen their grains and provide you with no help when you're trying to get a smooth finish on an auto repair job.

Some solders have no plastic range. Such solders are never suitable where capillary action is required to make your joint.

Solder Alloys

Soft solders are basically compounds of tin and lead in varying percentages. You can alloy in different metals. In some cases, as with lead-silver solder, there may even be no tin in the solder. Solders will usually have the alloy listed more clearly, with letters following numbers. As an example, a lead-silver solder with 3 percent silver would be a 3S solder. If the solder has at least 35 percent tin and also has more than 0.12 percent *antimony*, the solder would be listed with an A after the percentage number telling you how much tin is in the material—35A for this example. If the antimony content of the solder rises to no more than .5 percent, and tin content remains at least 35 percent, then the letter B follows—35B. If the letter C follows the number, then the solder will have no more than 6 percent antimony (as compared to the tin) and from 25 to 40 percent of tin. Assuming a B type solder, with 40 percent tin, the designation would be 40B, meaning that there is 40 percent tin, .5 percent antimony, and 59.5 percent lead in the compound.

Solders containing too much lead do not usually form good joints because lead doesn't diffuse well on brass. Too much tin, though, will make the solder brittle. A sufficient tin content will add greatly to the workability of the solder, so that 50-50 solders are among the easiest to use, and 60-40 (60 percent tin) are even easier to work. When tin content drops below 35 percent, the solder becomes harder to work. When it drops down as low as 30 percent, even the most expert solderers are likely to have trouble using the solder and getting a good joint.

Various other metals are added in order to give the solder certain characteristics important in special applications. Antimony provides extra strength, if used in miniscule amounts, and will also increase electrical conductivity. Silver improves wetting characteristics of the solder (wetting is simply the ability of the metal to flow after reaching melting or liquidus temperature) and is also an aid to electrical conductivity. Other characteristics that allow superior joints in dissimilar metals are supplied by cadmium, silver, and zinc together in lead. *Bismuth* may be added to lead-tin alloys to drop the melting temperature well below what is normal for even soft solders.

Soldering Fluxes

After you give the parts to be soldered a thorough mechanical cleaning, chemical cleaning to remove oxides will still be needed, just as it is with silver soldering and braze welding. The remaining oxides are removed with fluxes, which also form a thin film over the work surface to prevent contact with air and the resulting formation of fresh oxides before the joint is finished. There are several fluxes used for soldering, with the two types most helpful to the do-it-yourself welder being corrosive and noncorrosive fluxes. Corrosive flux may contain a compound such as zinc chloride and must never be used on electrical or electronic connections. The flux action will continue after soldering is completed and will eventually cause a high resistance condition in the joint. Noncorrosive fluxes are essential to electrical and electronic soldering jobs.

Noncorrosive fluxes generally have a weaker oxide removing action than corrosive fluxes do. They are not as stable over a wide range of melting temperatures as the corrosive fluxes are. The higher the melting temperature of the solder you are using, the more likely you are to need a corrosive flux for proper bonding. Make sure you remove all flux residue immediately after soldering. Flux conditions require you to use a relatively low temperature solder for all electrical and electronic work.

Some metals require more specialized fluxes. Metals needing such special fluxes include aluminum, cast iron, magnesium, stainless steel, and a few others. You'll be able to pick up fluxes for any of these metals from your supplier. If the hardware store where you've probably bought the rest of your soldering equipment doesn't carry them, many welding distributors and some electronics supply houses often will. While a slapdash job can be done on these metals if you use standard fluxes, the special fluxes are needed for best results.

Fluxes generally come in two forms. Some are pastes that you brush on; others are liquids that come in squeeze bottles. You will also find flux-cored solder wire on small reels.

Soldering Torches

While large soldering irons are available, most nonelectronic soldering can easily be handled with a torch. Most single gas soldering torches come with a fair range of tips—a standard pencil flame style, a fan-shaped tip, and a solid tip meant solely for soldering. You have a choice of using propane or MAPP gas for the soldering job, though in almost every case the propane gas will be cheaper and more efficient. The higher temperatures produced by MAPP gas are seldom needed for soft soldering (Figs. 10-10 through 10-12).

The wide range of single gas torches available today makes a selection somewhat difficult. Cleanweld has a pressure regulated single gas torch that is very precise, but it does cost more than a standard unregulated propane torch. Several companies offer propane torches with pilot lights and squeeze type handles. When extensive soldering needs to be done, this pilot light feature might save you a considerable amount of fuel gas, if the torch you generally use has to be left burning while you go from one joint to another and set up the joints, or a lot of hassling with a spark lighter.

Oxyfuel gas torches can also be used for soft soldering, though propane is usually the fuel gas preferred over acetylene or MAPP gas. Acetylene and MAPP gas produce flames too hot for almost all soft soldering, causing great difficulty in getting really good solder joints. Special heating and soldering tips are available from oxyfuel gas manufacturers. You should never attempt to soft solder with a welding tip on any oxyfuel gas torch.

Gasoline-operated torches seldom are used and seen today. There are plenty of used ones around, but I would advise against using one. Replacement parts are hard to find, and the fuel is exceptionally unsafe. If you want to check one out for its antique value, look for the oldest models possible. They are more heavily constructed of more expensive alloys than are readily affordable today.

Soldering irons will be more extensively covered under electronic soldering, but basically any soldering iron is a

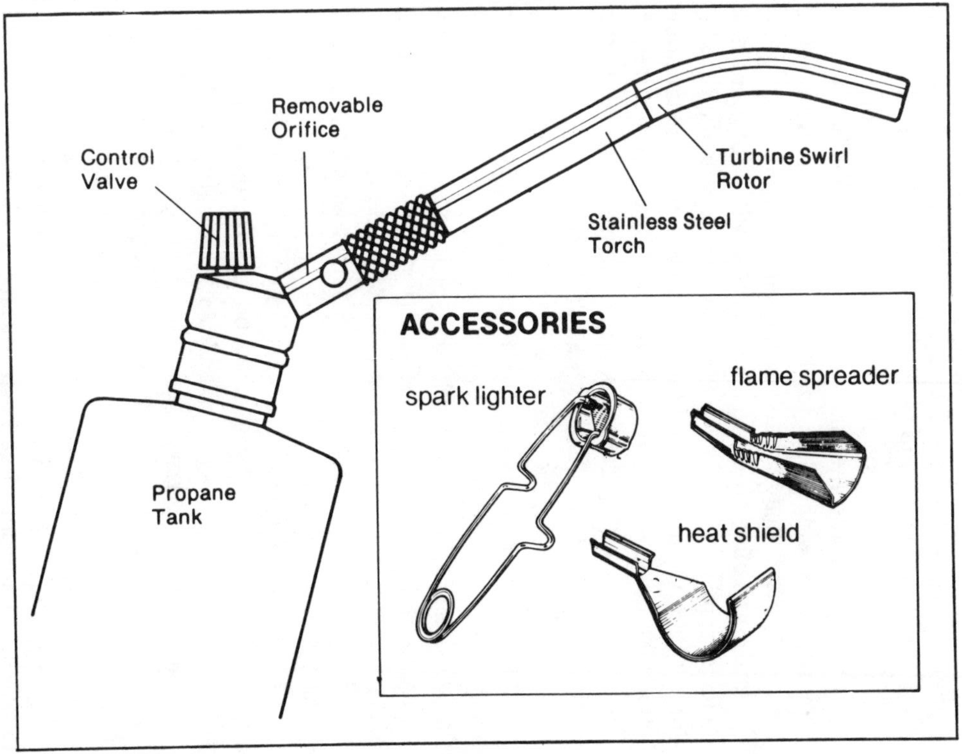

Fig. 10-10. A model 300 series propane (courtesy of Wingaersheek Inc.).

Turbine Swirl Rotor

Stainless Steel Torch

Removable Orifice

Control Valve

Propane Tank

ACCESSORIES

spark lighter

flame spreader

heat shield

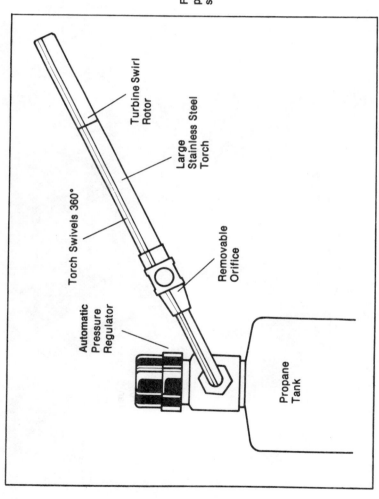

Fig. 10-11. A model 400 series propane torch (courtesy of Wingaer-sheek Inc.).

Turbine Swirl Rotor

Large Stainless Steel Torch

Torch Swivels 360°

Removable Orifice

Automatic Pressure Regulator

Propane Tank

219

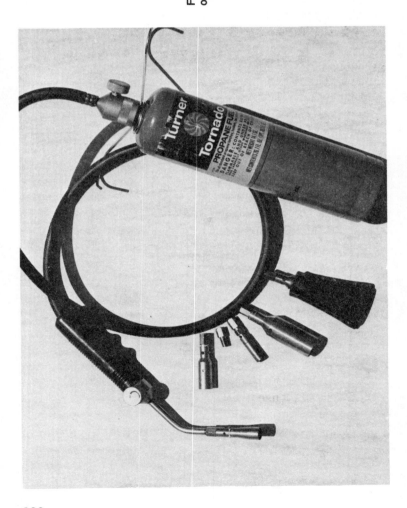

Fig. 10-12. Propane torch and various tips.

piece of metal, usually copper, drawn to a point and attached to a handle. Heat can be supplied to the iron's tip either by an internal electrical resistance circuit or by application from a handheld torch. Virtually all modern soldering irons are electrical.

Torch Soldering

The cleaning of surfaces is the most important step in any soldering job. Most soldering involves making joints in copper pipe runs. While mechanical cleaning is of great importance, you must not open up the prefitted joints to the point where the solder won't flow in and remain to seal the joint. Generally, you'll find that a good wiping with a fine grade of steel wool will suffice on pipe ends and on the interior of the pipe fitting. The interior of the pipe, after cutting, should be reamed to cut down on water flow restrictions caused by any burrs the pipe cutting may have left. A pipe reamer is built into the back of your tubing cutter; simply swing it out and rotate it gently inside the pipe to remove any burrs. Use your steel wool to clean both the inside and outside of the joint. Both surfaces to be joined must be bright and shiny. With copper pipe, a chemical cleaning with a commercial solvent before the flux is added will remove any grease or oil that might have been picked up in transit or storage. Remove the solvent from the pipe with hot water and flux the joint.

Heat the assembled joint until you see the flux starting to bubble. Apply solder on the side of the joint away from the direct flame of your torch. If you are making a pipe joint, move the solder and the torch around the joint until you can see a line of solder all the way around. You have made a sweat soldered joint in basically the same manner as you sweat solder a silver soldered joint (Fig. 10-13).

If you can not use this technique, there is another method of sweat soldering that you can substitute. Perform the cleaning and fluxing jobs. Next, simply precoat (tin) the pieces of pipe before assembling, using a very thin coat of solder floated onto the surfaces to be joined. Add a bit more flux to the tinned sections and assemble the joint. Apply heat to the

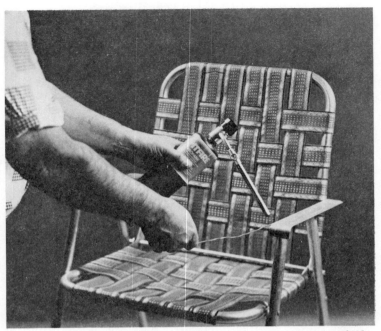

Fig. 10-13. Light metals solder easily with propane torches (courtesy of Wingaersheek Inc.).

exterior surfaces of the joint until the solder begins to flow, and you see a silvery line at the joint's exterior edge. This tinning method of sweat soldering is not as frequently used as the simpler flow-on method first described. It is a bit more complex, and it may even force you to file down some of the tinning solder to allow the joint to fit together properly. You'll find it more difficult to tell exactly when you've got a well-sealed joint. Still, tinning is useful for vertical pipe joints and is also useful when you're assembling jewelry and other flat items. Often with copper pipe you'll find it simpler to depend on capillary action to carry the required amount of solder into the joint to provide a seal. Given careful preparation, you should have no difficulty in getting a good joint every time.

Joint Designs

As with silver soldering, you can use lap and scarfed butt joints for soft soldering, but there is one major proviso. If any shear stress is going to be applied to either form of joint when soft solder is used, you must also supply the joint with some form of mechanical support such as screws and bolts or tie-down straps. Such support is essential so that a soft soldered joint remains intact under stress. Even with good mechanical stress, you'll find it best to avoid butt joints altogether, as they separate far too easily at the slightest pull (Fig. 10-14).

Sheet metal sections with soft solder joints can use a variety of joint designs. Small metal brakes are readily available to help you make these joints. If the brake is small enough, the price won't usually be totally out of sight. Generally, you'll need to use soft solder with such sheet metal work only when the joint must be sealed tightly enough to keep air or liquid from penetrating it.

Clearances for soft soldered joints are the same as for those which are silver soldered—0.003 to 0.006 inches for copper and about double that for aluminum. The base metal is also the portion of the work that must be heated, not the solder. First, flux the joint. Then apply the flame from the torch directly on the joint. When the flux starts to bubble, feed the solder to the joint. Once your joint shows a line of solder all the way around, remove the torch and let the joint cool to room temperature. As the joint is cooling, do not move it at all. If you can't touch the immediate area of the soldered joint with your hand comfortably, then the joint is still too warm to be moved without damage to the bond.

If possible, you should plug the ends of any pipe you are soldering to cut down on heat loss. Copper conducts heat extremely well. Aluminum has much the same rapid heat

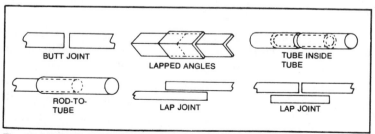

Fig. 10-14. Solder joint designs (courtesy of Wingaersheek Inc.).

dissipating property, so you may find that more intense heat is needed to get the solder to melt all around or along a joint. Aluminum requires a special flux, while most of copper alloys can be fluxed with almost all the available types. Pure copper, if you locate any, should be fluxed only with noncorrosive fluxes to prevent oxidation of the metal. Aluminum fluxes are extremely corrosive because they are especially compounded to fight off aluminum's naturally rapid oxide film buildup. Fluorides are essential ingredients of effective aluminum fluxes. You will have to make sure there is adequate ventilation, as the fumes are harmful.

Soldering Cast Iron

Getting a decent soldered joint is difficult because the cast iron resists the wetting action of solders. Cast iron is one of the most difficult metals to solder effectively. First, the metal must be thoroughly cleaned, both mechanically and chemically, of all rust, scale, and grease or oil. You must have a surface that is bright and shining before you add flux. Then use a strong, corrosive flux and a fairly large heat source (the size of the heat source depends on the size of the piece being soldered). If no oxyfuel gas equipment is on hand, you can use several single gas torches to bring the material up to soldering temperature.

If the suggestions regarding cleanliness and extra heat are followed, you will get good solder joints in cast iron. Stay with standard heat sources, follow the other suggestions, and there should be no soldering jobs you can't carry out until you decide to work on electronic items.

SOLDERING ELECTRONIC COMPONENTS

Soft soldering electronic components and electrical circuitry presents you with a somewhat different set of problems than does soldering pipe or sheet metals. Joints in electronic kits need less strength because they are predesigned not to have loads or stresses placed on them, or they are designed to resist such loads or stresses with little help from the solder. These joints, whether to hold a circuit board in place or to connect a terminal, will often be located next to items (such as transistors) that are exceptionally sensitive to heat and often easily destroyed by moderate amounts of heat.

This heat overload possibility brings in major tool changes you must make for electronic soldering. Under no circumstances can you possibly use the kind of soldering tool that you would use on even the lightest copper pipe or tubing.

Remember that soldered joints in electronic gear, whether it be a fancy 200-watt stereo amplifier or a speaker kit, must do a good job of conducting electricity. This need brings about a requirement for special solder alloys and near perfect solder joints. You must control the amount of solder going onto the joints, for great globs of solder could easily bridge between circuits and cause problems ranging from interference to short circuits that might result in major damage to the gear you're assembling. Neat is absolutely mandatory in electronics soldering.

Solders for Electronic Work

The plastic ranges of solders can vary quite widely. One of the first requirements for solder used in electronic work is a short or nonexistent plastic range. As an example, SN30 solder offers you a plastic range of a full 135 degrees Fahrenheit (57 degrees centigrade). Type SN63, a solder extensively used for electronic component soldering, offers exactly no plastic range. Obviously, then, your preferred solder for electronic work is one that solidifies almost the instant you remove the heat source.

Type SN63 offers another desirable advantage for electronics work—a very low melting point of 361 degrees Fahrenheit (183 degrees centigrade). This low melting temperature reduces the possibilities of heat damage to components being soldered or to those nearby. The very narrow—here nonexistent—plastic range means you'll have less problem with a phenomenon known as cold solder joints (which are just as commonly called fractured solder joints). A *cold* or *fractured solder joint* is one where movement before the solder sets causes a poor circuit. These can be tough to trace later on.

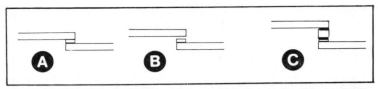

Fig. 10-15. Bad electronic soldering joints. The three types of "cold" solder joints. (A) A fracture joint. (B) A true cold solder joint with too little heat used. (C) A joint that hasn't been properly cleaned.

There are three kinds of bad solder joints, all grouped under the name cold solder joints, and each has a different cause. One is the fractured joint. A second is a true cold solder joint where too little heat was used to give proper bonding. The third is a solder joint made either without flux or proper cleaning. Any of these conditions will present you with major problems in electronic assemblies. A bad solder joint will cause higher resistance than the circuit was designed to accommodate and will interfere with the performance of the electronic equipment (Fig. 10-15).

When the solder you use has no, or almost no, plastic range, the solidification at the joint is very rapid. You can forget some of the excess care needed to correctly assemble the parts in a kit or to later make repairs.

Fluxes

For electronic work, only rosin fluxes can be used. Rosin flux, whether in a bottle or in the core of the solder, is the only form of flux that will not attack the metal through corrosion. Such corrosion will later result in a faulty solder joint. Never use an acid or other corrosive flux, no matter how mild its action is claimed to be.

Soldering Guns and Pencils

Soldering pencils for electronic assemblies are essentially the same as the larger soldering irons briefly described earlier in the chapter. There are several differences—the most obvious being the size of the tool. Soldering pencils are exceptionally small, often not much bigger than a large fountain pen. The soldering pencil's wattage, compared to full size soldering irons, is miniscule. A 15-watt soldering pencil is just about the minimum useful size for any kind of work. A 25 to 30-watt pencil will offer you the greatest utility when the component size varies, and you wish to spend most of an evening working on a kit. For heavy-duty electronic soldering, you may want to move up to a 40 to 50-watt soldering pencil, though that's about the maximum size for kit work.

Many soldering tools for electronic work are available. Some of these temperature and phase controlled tools are required for production line jobs, but the increase in cost compared to the basic soldering pencil makes such equipment a rather poor choice for any hobbyist solderer.

Such tools are frequently called soldering and desoldering stations. They come with everything from a stand for the pencil to a control unit for the temperature and the vacuum (for use when removing solder during desoldering). Ungar, Weller, Edsyn, and others produce several versions of these stations, along with soldering pencils. These units are safe to use and offer such features as nonmeltable cords, three-wire grounding plugs, and many tips suitable for various electronic soldering purposes.

Generally, most soldering guns will offer 100 watts as their lowest heat setting. These soldering guns are suitable for heavier electronic work, but they are difficult to use on printed circuit boards and in other situations where the work may be crowded in and where components that are easily damaged by heat may be located. Tip design is also very limited.

Soldering Pencil Tips

The tips for soldering pencils are generally made of copper, though some are iron plated to increase their life span. Copper eventually erodes because of the solvent action of the solder. Copper tips may be refiled to shape when erosion makes their shape unsuitable to the jobs you are doing. Allow the pencil to cool to room temperature before you gently file the tip back to shape.

Before use, copper tips must be tinned and cleaned of contaminants that form during use and storage. Before you plug the pencil in, wrap several loops of solder around the tip

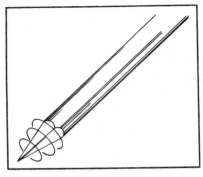

Fig. 10-16. After cleaning, wrap the iron's tip with solder and heat to get proper tinning.

(Fig. 10-16). Plug it in and let the solder melt. Finally, use a damp sponge to spread the solder evenly over the tip, and you're ready to start soldering. During soldering, contaminants will build up on the tip. For a normal buildup, simply wiping the tip across a damp sponges will remove the foreign matter (Fig. 10-17). When you're using a copper tip, you'll find any heavy buildup that forms can be easily removed with a small wire brush. After such cleanings, you should wipe the tip with a damp sponge and retin it. Iron plated tips will not need wire brushing and will actually be ruined if a wire brush is used.

Tip Styles

There are three basic soldering pencil tip styles. The heavy spade tip is used for larger grounding straps, bus bars, and large terminals (large is a relative term in this instance because electronic soldering seldom uses wire sizes above number 18). The long spade tip is used in moderately restricted areas in television sets, stereos, and other relatively large components. For transistorized circuitry in very small components, and for other areas where parts are extremely heat sensitive, you will use the minispade tip. You will use the minispade tip on printed circuit boards (Fig. 10-18).

The Ungar soldering pencil in Fig. 10-19 doesn't offer a large or heavy spade tip, but instead has a triangular tip. In addition, there are several heating unit variations to fit into the basic pencil handle. My three elements offer 27 watts (600 to 650 degrees Fahrenheit—316 to 344 degrees centi-

DAMP SPONGE

Fig. 10-17. Use a damp sponge to wipe off light buildup as you work.

grade), 42 watts (700 to 750 degrees Fahrenheit—371 to 399 degrees centigrade), and 50 watts (800 to 850 degrees Fahrenheit—427 to 455 degrees centigrade). When such interchangeable heat elements are used, you should coat the threads with an antiseize compound before inserting them in the handle (Fig. 10-20). The compound keeps the heat from locking the parts together and making further changes difficult or impossible.

While soldering pencils with interchangeable heating elements are more expensive than the noninterchangeable types, you get much greater versatility and extended tool life because you can replace a heating element or tip if one should wear out. This saves you from having to buy soldering pencils in several heat ranges.

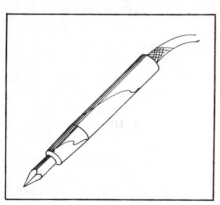

Fig. 10-18. Minispade tip.

229

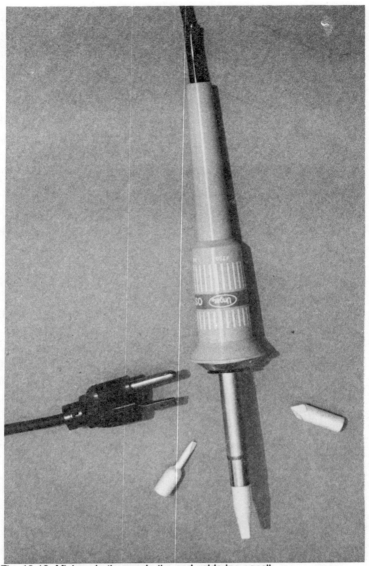
Fig. 10-19. Minispade tip, spade tip, and soldering pencil.

When soldering electronic components, cleanliness is of absolute and utmost importance. Dirt, oil, and grease are all easily removed with alcohol, but the use of flux is still essential to a good bond. Rosin core solder is most commonly used for this purpose when you solder electronic components. The core in the wire of solder melts, lifting off the oxides and floating them to the surface of the joint, where removal is a simple matter. In cases where you find a heavy buildup of oxides, you will have to use a liquid or paste rosin flux to remove such heavier deposits.

You first add solder to the tip of your soldering pencil and allow a small drop of solder to form. At this point your application procedure may differ slightly, depending on the type of component being soldered. If the solder doesn't flow

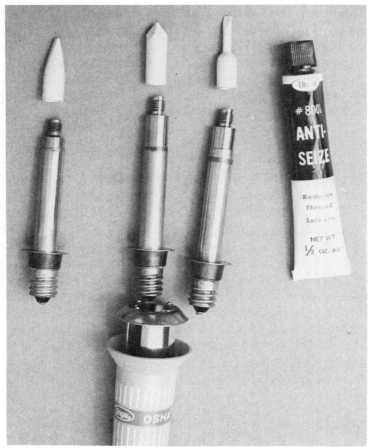

Fig. 10-20. Soldering pencil with different heating elements, tips, and antiseize compound.

upwards from capillary action when you are soldering a component from the bottom of a circuit board, you must move the pencil tip to the top of the board. Apply wire solder from below to make the joint. In this way you will force the solder to flow to the heat source by capillary action, thus getting a good joint.

Timing is important when soldering electronic connections. The time you need to make any particular joint will depend on the size of that joint, the wattage range used, and the tip size. Taking two to three seconds for each solder joint is about ideal. When you're making a series of connections in rapid succession, the pencil tip will cool down to where it cannot supply the required heat after a few joints are made. In such cases, it is good practice to give the tip time to attain proper heat after every dozen or so joints are made. The longer you have to apply any heat to a joint, the more heat will be transferred to the component, thereby increasing the chance the component will suffer heat damage. Most electronic components will be damaged if the internal temperatures reach no more than 350 degrees Fahrenheit (177 degrees centigrade).

When you check the manufacturer's specifications for components, they almost always include heat limits. Most small diodes, transistors, capacitors, and integrated circuits can withstand heat up to about 392 degrees Fahrenheit (200 degrees centigrade). Locating a heat-damaged component after assembly can be quite a complicated chore.

Whenever you think components are likely to be damaged when you solder them, you should use small *alligator clips* as heat sinks. A heat sink will absorb the heat that might otherwise reach the component, sparing it possible damage. The shorter the lead on the part you are soldering, the more susceptible to heat damage that component will be, and the more necessary you will find the use of a heat sink (Fig. 10-21).

Desoldering

Sooner or later you will mess up a solder joint, and sometimes older gear will need a few solder joints removed

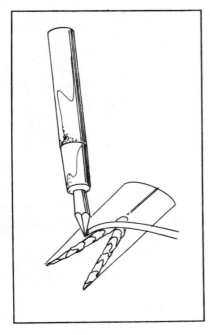

Fig. 10-21. An alligator clip used as a heat sink.

so components can be replaced. Obvious solder joint faults can be seen or felt. Any solder joint in which the connector can be moved with your fingers or with a very gentle tug from a pair of needlenose pliers is bad and must be redone.

When faulty joints are located or repairs are needed because components have failed, *desoldering* is essential. It's possible to heat the lead and then just lift the component out of the circuit, but you then have a problem with the old solder left behind. This old solder will make the insertion of a new component very difficult. While you can buy elaborate desoldering irons with built-in vacuum lines, a standard soldering pencil with a bulb style screw-in desoldering tool will work just as well, though more slowly (Fig. 10-22). Other kinds of desoldering tools are also available at reasonable prices. Edsyn offers a unit called the Soldapullt and another called the Soldavac. Both have spring-loaded plungers with trigger buttons allowing you to activate the suction once the old solder is molten. Such tools work very well, but a conventional bulb type does, too, and is a bit cheaper.

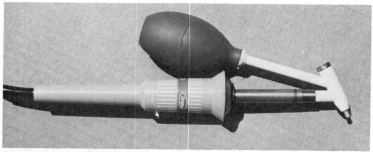

Fig. 10-22. Bulb style desolderer.

Tips

Most component leads will have to be bent before you can insert them in the circuit. Use a pair of long-nosed pliers for the purpose to avoid damaging the components.

Make sure your wire cutter blades are sharp. A neatly trimmed lead is easier to insert in the circuit than one with burrs.

Because stranded wire separates easily, you should always tin the leads with such wire before inserting them into the circuit board. Make sure, though, that you do the tinning carefully. Use just enough solder to hold the wire strands together. A blob of solder on the end will make the wire as hard to insert as it would be with the separated strands going every which way.

When you have components to be soldered into a printed circuit board and find the circuit board holes oversized, you can hold the components in place with double backed tape while carrying out the soldering process.

Swaged terminals will need some solder run between the terminal and the circuit pad. Treating the terminals in this manner ensures immediate conductivity and helps to prevent corrosion.

Accessories

A soldering pencil, some tips, and a desoldering attachment are not the only tools required for electronic soldering. A pair of needlenose pliers of medium length and a pair of miniature wire cutters are needed (Fig. 10-23).

234

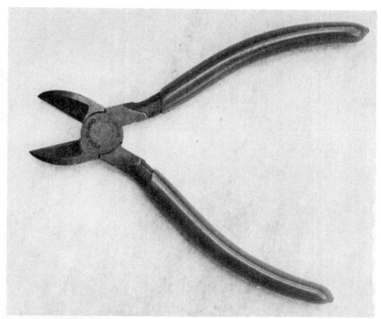

Fig. 10-23. Wire cutters.

You will also find a pair of long, slender tweezers helpful when you go to insert leads from components into holes in the printed circuit boards. The tweezers are also handy for removing desoldered pieces.

Alligator clips with fine teeth will make excellent heat sinks. You will seldom need more than a single clip at any one joint. If you have six clips or more on hand, you can prepare joints well ahead of your soldering pace.

You will also need a small brass wire brush with fine bristles for cleaning soldering pencil tips of copper. A small sponge kept damp during soldering is good for wiping off the lighter contaminants and excess solder.

A soldering pencil holder is not essential, but it is helpful. I have a wire basket type (Fig. 10-24). The holder is where your pencil rests every time you sit down to work. Your holder will also keep the pencil from rolling around and disturbing some almost completed work. The basket type is generally the least expensive. Make sure the base sits at a good angle and is made of fairly heavy metal.

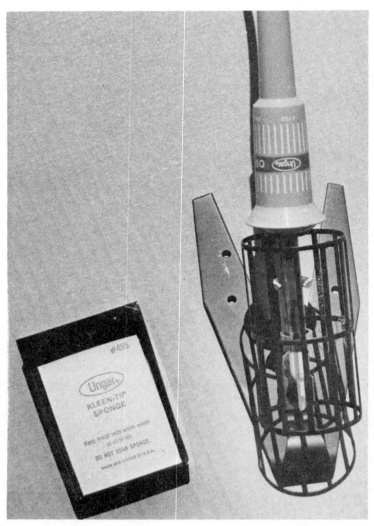

Fig. 10-24. Basket style pencil holder.

Safety

Soldering requires attention to safety measures. Obviously, no matter the heat source, you should never forget that you're working with materials hot enough to do plenty of harm to the human body. Keeping the heat away from you is paramount.

Molten solder seldom spatters as widely as molten welding alloys. Dripping is always possible, so take care.

Safe handling of small single gas torches is about the same as for any torch. Keep the flame away from you. Also, respect those small, disposable propane and MAPP gas bottles. Don't bang them around. Protect them as much as possible from extremes of heat and cold.

Use only soldering guns, irons, and pencils that are Underwriters Laboratories (UL) listed, double insulated, or fit a grounded, three-wire outlet. If your home or farm only has a two-wire set of circuits, use an adapter on grounded soldering tools. Make sure to ground the adapter to the receptacle's plate screw.

Rosin fluxes give off plenty of smoke, but there is nothing in the fumes that is known to be harmful. The smell isn't pleasant, though, so provide adequate ventilation. When soldering, good ventilation is needed for another reason. The fumes given off by the lead in the solder, if the material is severely overheated, can be toxic. If no ventilation is possible, be sure to use a respirator.

Chemicals used for cleaning surfaces to be soldered can also be harmful. Alcohol, frequently used for cleaning connections for electronic soldering, is also flammable and must be handled with care. Any acids used for cleaning surfaces should be kept in appropriate containers and used with extreme care, particularly since some of the stronger acids can cause severe chemical burns.

Zinc chloride may irritate the skin and cause dermatitis. Be careful how you handle it, and be sure to wash your hands after any contact with it.

Using carbon tetrachloride is now illegal in many states, and the substance is best not used at all. It is a very dangerous material, particularly when heated.

Chapter 11
Container Welding

You may have to repair containers like large drums, heating fuel tanks, gasoline tanks, and other storage tanks. If a container has never held any flammable or volatile materials, the repair job is one you may perform in relative safety with not much additional preparation.

Each year, though, many people are injured, including many professional welders, while working with a torch or arc welder on containers ranging in size from huge oil and gasoline tankers to small 2-gallon gas cans for lawn mower and chain saw refueling. In virtually every case of injury, the explosion and resulting injury is the fault of the welder using techniques that don't purge the container of any volatile fumes present. Though you may empty a gasoline tank before starting to weld it, that tank is not really empty. It will contain vapor from the gasoline, a highly *volatile* substance (volatility is the measure of a substance's ability to explode or burn rapidly). Many substances normally not considered very volatile become extremely so when subjected to the intense heat of welding or brazing operations, so that a tank which has contained motor oil, heating oil, tar, oil-base paints, many types of plastic, and some kinds of rubber compounds, plus dust-producing compounds even in small amounts, may be dangerous for you to work on. Container explosions almost always result in severe injury or death, so care is needed.

Remember that even substances painted on metal can give off fumes which could prove volatile under extreme heat.

CHECKING WELDS AND REMOVING EXPLOSION ELEMENTS

The first procedure before attempting to weld, braze, or cut any container, though, doesn't involve materials held in the container or painted on it. You must first look at the welds that the original maker of the container used to hold it together and seal its seams. Check immediately to see what sort of weld joint was used to produce those seams. If lap joints, and most particularly double lap joints, have been used, discard the container and replace it with another (Fig. 11-1). Because a very slight weld porosity can allow volatile substances to penetrate any weld, lap seam welds are dangerous because the lap seam may hold such penetrating volatiles in the lap for quite a long time, sometimes several years past the time when the container actually held any volatile material. You should never attempt to weld any container with lap seams. Replace any containers with lap seams before any rupture occurs.

If your seam check turns out all right, with butt joints being used in the seam welds, you can remove one of the three elements needed to cause an explosion. Oxygen is one basic element. Ignition is another, and combustible material is the third. It seems impossible to imagine any chance of a welder removing the ignition source because welding is always accompanied by temperatures in excess of the ignition or flash point of any volatile material. You are left with the other two elements to be removed from the picture in order to safely weld containers. Removing either one will do the job, but eliminating both prevents any possibility of explosive mishaps.

MANDATORY STEPS

The following steps should be mandatory for welding, cutting, or brazing operations on all containers that you know have held volatile materials. On containers that have held unknown materials, it's wise to follow the same procedures.

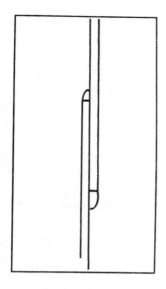

Fig. 11-1. A double lap joint is not good for use with containers that may carry volatile materials.

First, you must drain the container as thoroughly as possible, and then flush it with cold water. In cases where there is fuel oil, use warm or hot water for the flush.

Locate someone who has a steam jenny or other steam gun setup. Completely steam clean the interior of the container at least twice. Make certain that the container has a reasonable cooling time between applications of steam. There is a theory that the first steam cleaning could produce static electricity in the still not quite purged container, allowing heat from the second steam cleaning to bring the static charge up to a point where it might discharge, tossing sparks and causing an explosion. Total steaming time for small containers should never be less than an hour, and you should add time liberally as the container increases in size. Make certain that the vent openings in the container are left open so that volatiles can be carried off in the steam (Fig. 11-2). If the container doesn't have a vent of some kind, you cannot weld it (small entry openings are not enough venting). Sometimes you may have to cut vents with a cold chisel, a procedure that can itself prove hazardous in some instances.

To steam purge containers the correct length of time, you can use a rule of thumb. For every container up to 200 gallons, steam clean for at least one hour. If the container is greater than 200 gallons, add an hour of steam cleaning time for each portion of 200 gallons that it is larger. Thus, you would steam clean a 190-gallon tank for an hour, while a 275-gallon tank would require two hours of cleaning time, and a 500-gallon tank would need three hours.

OPTIONS WITH COMBUSTIBLES

After steam purging, a container may still have combustibles present. If so, you have several options. First, you can try to borrow an explosive meter from your local gas utility and use it to make a check. If that isn't possible, use an inert gas to shield any possible volatiles from contact with oxygen. Water can also be used as a shield, but the inert gases usually do a better job most of the time. Argon, helium, nitrogen, and carbon dioxide are all useful, though you should avoid carbon dioxide when you weld ferrous metals.

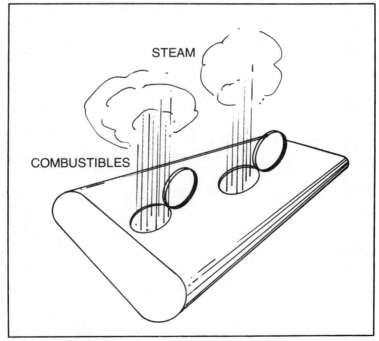

Fig. 11-2. Steam cleaning a container.

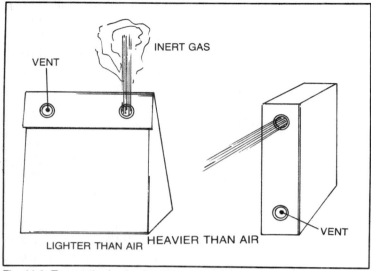

Fig. 11-3. Two methods of using inert gas to purge containers.

If volatiles lighter than air are in your container, a heavy inert gas such as argon can be run in through a top opening, which will force most of the volatiles out of any other top vent (Fig. 11-3). For combustibles that start out heavier than air, like many fuel and motor oils, helium is injected into the container from the bottom or from a *very* low side vent. Containers small enough for you to turn can easily be moved around, but you might need to add a vent or increase vent size with a chisel. When you use a chisel to cut vent openings in a container, you must have that container partly full—at or a bit past the spot to be cut open—of inert gas or water to protect yourself from an explosion in case the chisel throws sparks as you make the cut. If the container metal is ferrous, the chisel will almost certainly throw sparks. The container can later be rotated so you can reach the newly cut vent and weld or braze is shut as needed (Fig. 11-4).

Water used as an agent to shield against combustibles is apt to be more popular around the workshop because of its lower cost compared to the inert gases. Start the purging job by steaming as usual, or you can use a caustic solution that is described shortly. When you're using water to displace the air

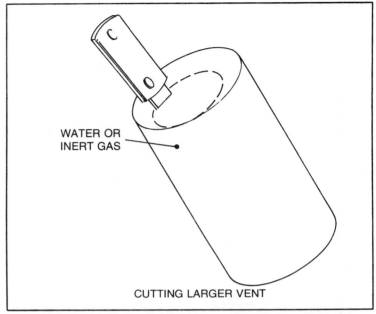

WATER OR
INERT GAS

CUTTING LARGER VENT

Fig. 11-4. Cutting a larger vent, using water or inert gas as a shield.

in a container, the section to be repaired must be placed as close to the top of the container as possible, obviously with the water level just below the rupture to keep the water from running out onto the torch flame or arc. Some small cracks may not leak much even when below the water line, but the water will aid heat dissipation and make welding or brazing very difficult (Fig. 11-5).

Vents for the water methods must be large enough to allow the free escape of air from the container being welded, and the air space above the water line must be as small as possible. Otherwise, there is still too great a danger of explosion.

If you are unable to use shield materials and can't borrow an explosive meter, another test can prove helpful in preventing explosions. After your thorough steam purge of the container, place the container so that there is a solid barrier wall—I prefer at least 8-inch thick concrete block—to resist an explosion. Tie a small single gas torch, preferably a cheap

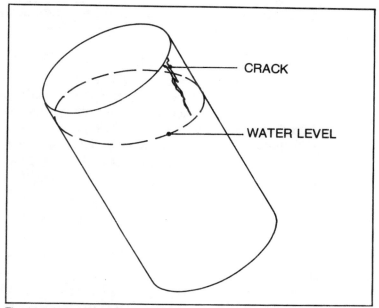

CRACK

WATER LEVEL

Fig. 11-5. Using water as a shield when preparing to weld a container.

one, to a closet pole at least 8 feet long. Get behind the wall, light the torch, and use the pole to move it out and around to the container. Insert the torch flame as far as possible into all openings in the container. This check will quickly serve to let you know if there is likely to be an explosion especially if, after testing all openings, you use the torch to heat the sides of the container and then do a retest. You may blow up the container, but you're behind a good, solid wall if it explodes.

CAUSTIC PURGES

There are several reasons why *caustic purging* is the least desirable way to purge a container. Using a caustic solution boiled in a container will clean out volatiles, but if the container is large, plenty of fuel and time must be expended to get the solution up to a boil and to keep it boiling. Caustics such as *lye* are dangerous. Bad burns are almost a certainty if even a drop of the cold lye solution gets on your skin, and the solution's caustic activity is increased markedly by heating. Steam is also likely to cause burn problems if not handled well, but steam jennies are reasonably safe to operate and have long wands to allow the operator to stay well away from the steam flow.

If you have to use a lye solution to boil out combustibles from a container, make certain you don't use one of the packaged drain cleaner products. Many of these contain aluminum chips or shavings which produce a hydrogen gas— another explosive. Use only pure lye. Take every precaution about getting the stuff on you or any piece of material you value. Do not carry out a caustic purge indoors, no matter how small the container or nasty the weather. Whenever possible, use steam for purges.

Chapter 12
The Welding Workshop

Setting up a welding workshop involves more than installing a torch and cylinders, or an arc welder, in a convenient corner. You must select a safe portion of the home to work in. A corner of the basement of garage *may* prove suitable, but you can also experience trouble in these areas if added protection isn't installed. Because of such safety needs, it's best to do all indoor welding in one spot. A windowless or fanless welding shop can provide great hazards to the lungs, even with substances not ordinarily considered hazardous when heated. Enclosed fumes from oxides are very hazardous, so ventilation is an important consideration.

VENTILATION

One man found a most desirable location just a day after he brought home a new Sears arc welder. The machine was set up in his garage, close to the front and on the concrete floor, as soon as the correct 60-ampere circuit was installed. (The circuit installation was simplified by the entrance panel being near that end of the house, and by the fact that he already had not only 200-ampere circuitry, but an auxiliary 100 ampere feed-in.) The walls of this garage are of concrete, and there are no windows. Whenever hot work is done, the garage door is pulled up. If there is no breeze, a 20-inch floor fan is used to blow fumes away. Usually there is enough breeze to dissipate the fumes.

Any indoor welding shop must have at least two fairly large windows for ventilation. If these windows don't provide enough air flow to keep fumes cleared on sultry days, you must install a fan to move the air. A temporary setup with a 20-inch floor fan will do the job when there are two windows open. When checking ventilation, you'll have to work without the fan to see if it is needed, and this work must be done with a respirator mask on at all times.

In cases where one or no windows are available in the selected workshop area, you will have to install a wall fan. If one window is available, a large, portable window fan may do the job. If no windows are available, you're going to have to cut through the wall, run wiring, and install the type of through-the-wall fan that has exterior louvers which can be closed to keep things warm when the shop isn't in use. Many auto body shop owners can tell you where to get these fans because they must use them in the paint room to clear fumes. Sizes and air movement capacities vary greatly, depending on the size of the room to be cleared (Fig. 12-1).

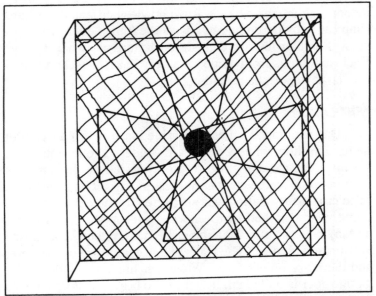

Fig. 12-1. Window or through-the-wall fans may be needed to properly ventilate an area used for welding.

OUTDOOR SETUP

There are many ways to set up gas welding equipment outdoors, but arc welding equipment not mounted on a trailer or pickup body is virtually impossible to use. This is simply because there is little power available. It is rather difficult to manhandle an arc welder of any capacity even though you may have the proper circuits run to a barn or garage exterior.

Setting up outdoors almost automatically eliminates any problems you're likely to have with ventilation. You can't store gas welding equipment outdoors in the weather and expect it to last a lifetime. Cylinders will rust, and hoses will rot. Gauges may fill with rainwater or dew. Welding rods must never be left out overnight. You must have a cart for your cylinders, hoses, and torches when you work outdoors. Like an arc welder, full sized cylinders are not easily moved around any distance without a good cart under them. Indoors, with one working area in use, the tanks can be chained to a wall or post. Outdoors, they must be mobile (Fig. 12-2).

Setting up the cylinders, installing gauges, and other jobs are easier outdoors than indoors. First, there is no need to worry about dispersion of gases when you clear the gauge seats or blow talcum from the hoses. Second, you're less likely to be bumping into things. Third, there is less danger of the first flame adjustments hitting a flammable object.

There are disadvantages of outdoor welding shops. First and foremost is probably inclement weather. Rain, snow, sleet, and high winds can ruin a day's welding. Cold weather is bothersome, but not as much as you might imagine.

SITING

Siting is a rather simple chore when you're doing outdoor welding. I've carted my Tote-Weld outfit many places to repair a piece of farm machinery, to weld bumper nuts onto a Jeep, and so on. With a Tote-Weld outfit, the weight is minimal—probably no more than 20 pounds.

Your outdoor workshop may be no more than a field somewhere on your property or on the land or other people who have asked you to do some work. You can build welding

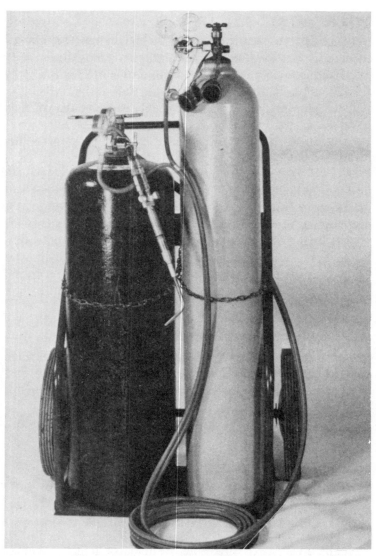

Fig. 12-2. Carts make tanks much more portable when they are used outdoors (courtesy of Airco Welding Products, Murray Hill, NJ).

benches and erect a special shed in which to store the equipment and metal supplies.

If no special shed is to be built, then consider the area where you hope to work and look around for the nearest storage area. I used to work out in front of my garage. Even with a concrete floor, the frame-attached garage wasn't a wise choice. It was filled with a lawn mower, weed cutter, chain saws, and a couple of motorcycles. That meant a trip of not much more than 15 feet with all the gear, and that is a consideration if you do much welding, especially if any of the welding is on heavier materials. You don't want to make several trips carting heavy gauge sheet metal over distances of 100 or more feet.

Carports as well as garage aprons can be handy spots for welding. The ventilation is almost as good as in the open air, and you are protected from the elements. The carport can also be used to store materials, but you must guard against theft.

PATIOS

While wooden decks are probably more popular than *patios,* sometimes a patio will be close enough to a storage area so that you can use it for welding. Because of their color, slate patios are less likely to show the stains associated with most heat worked metals (both are gray).

Bare ground is also suitable for welding, cutting, and brazing. Note that small amounts of molten slag can start quite a grass fire. Check for leaves and other debris when doing field welding. If you have to weld something on a lawn, don't expect the grass under the job to be around after you're finished unless it is protected.

BUILDINGS

Outbuildings around farms don't commonly have the electrical circuitry needed for arc welders, and the distance from the main service panel is usually too great to allow such circuitry to be run with any economy. Thus, it's back to gas welding equipment. If the outbuilding is basically suitable, the setup is much the same as for any indoor shop. You need good ventilation, a dirt, stone, or concrete floor, and either nonflammable walls or some form of protection for those walls. In the South, most outbuildings appear to be of frame construction, but in many states stone, or even brick, smokehouses and other outbuildings were often built on older farms. Such places can be ideal, once you install proper ventilating fans. Most buildings have very poor ventilation, so care must be taken.

Any work in or around outbuildings or areas with dried grass, straw, or leaves must be accompanied by the appropriate fire protection methods. You may be able to soak the area with the garden hose if it is close to the house. If you are away from water outlets, you should set out several pails of water, the number depending on the size and type of job. Large cutting jobs obviously require the most water. In addition, I would keep at least a 5-pound dry chemical fir extinguisher on hand if there are any flammable parts on the workpiece. Dry chemical materials won't work too well on spreading grass fires. If you act quickly, a pail or two of water will control the flames.

INDOOR SETUP

Whether inside a basement, garage, or outbuilding, there are some rules to follow. While the fumes from welding mild steel are not poisonous, a concentration of such fumes can readily upset your stomach and make your lungs ache badly. The more extensive the welding jobs you anticipate doing, the more intensive becomes your need for adequate ventilation. Welding for 10 or 15 minutes will seldom fill a reasonably sized shop with fumes, but an hour of welding may do so. If the material being welded is coated with zinc, as are galvanized steels and irons, you must wear a respirator mask.

Once the ventilation situation is determined, you can consider the other needs of your indoor workshop. First is the floor. Under no circumstances can a wood or plastic tile floor be suitable for welding and cutting. Many people feel that such floors can be made safe by installing a fireproof or fire-resistant coating, but the idea makes me nervous anyway. It takes only one white-hot drop of metal slipping through or over the edges of such protection to start a fire that may not be noticed until it is too late. Select a welding workshop area with a floor of concrete, stone, brick, or dirt.

Look over the walls. If they are frame walls, no matter the finish or lack of finish, you'll need some protection for them any time you work closer than 10 feet. Babcock & Wilcox (P.O. Box 923, Atlanta, GA 30393) has recently developed a material known as *Kaowool* which is a spun ceramic and can be obtained in ropes, millboard, batts, and so on.

Once the floors, walls, and ventilation are inspected, check for ease of entry. Metal and gas cylinders are not light. Even with a cart, plate and sheet metal can be tough to move if you have to go up or down a couple flights of stairs. This problem will occur mostly if you have a basement or outbuilding shop and can be difficult to solve if you have no other space to use as a workshop.

Setting up a welding workshop for your home or farm isn't a good idea when the equipment is in a basement. You must either plan on doing only lightweight work or provide an easy way to move materials, gas supplies, and finished products.

STORAGE

When considering your welding workshop needs, give some thought to the sizes of jobs you'll be doing. Storage needs are directly related to the number and size of the finished jobs you'll be turning out. Working with metal, you'll seldom be faced with the requirements of the home woodworker who decides to build a houseful of furniture, particularly since the metal is more stable in temperature and doesn't need a lot of storage time in the area where it will be worked. Metal scrap and extra materials will accumulate over time, and you'll need space for everything. As time passes, you'll probably find yourself more reluctant to toss away even small scraps, for even those may prove useful later.

You may find yourself adding hand grinding and drilling tools, more types of welding rods and brazing rods, brazing fluxes, and other items. There are many ways of storing the various items in a welding shop, and some seem rather casual. A professional welder that I know tosses most of his mild steel and other metals in a field near his home. The grass grows over them, and they rust away when not used. It is almost impossible to keep rust off mild steel when it's stored for any length of time, but I prefer at least a covered environment, even if it is nothing more than a sheet of plastic. Storing such materials outdoors when the quantities are large is often necessary. If you place some old pallets under the materials, the oxidation problem is reduced.

Shelves can be bought or built for storage of tools, fluxes, noncoated welding and brazing rods, and smaller pieces of new and scrap metals. Store plate and heavy sheet metal on edge against a clear wall. Place a board under the edge and in contact with the floor. Consider a large, heavy wooden box or a metal trash can for storing small and odd-shaped scrap pieces.

Coated welding and brazing rods or electrodes are best stored in a more controlled environment. The rods must be kept dry. Probably the best way to store the rods and electrodes is to locate an old, nonoperating refrigerator and place them inside. Fit some sort of lock to the door of any refrigerator that is not in constant use.

LIGHTING

Lighting may not seem too important in a welding workshop, whether for arc or gas welding, as the flame or arc provides plenty of light. You must wear goggles or a face shield to protect your eyes from the glare. Good light is necessary for reading plans and laying out the work on pieces to be cut, welded, brazed, and so on. Marks on metal tend to be fairly dim anyway, and poor light adds to the difficulty. Additionally, any areas where tools like grinders and power hacksaws are used need to be lighted well.

You need one good light over the plan bench and marking area, with another light over the welding and cutting area. Individual smaller lights should be placed over other work areas. Probably your most economical bet is the fluorescent strip lights readily available from most hardware stores. This lighting is available in single 2-foot, single 4-foot and double 2-foot and 4-foot strips in both sizes and is reasonably simple to install. A few of the smaller sizes can be plugged into wall outlets. Install plenty of outlets in the welding shop, and make sure that all the 120-volt outlets are set up with 20-ampere circuits. Much depends on the type of welding shop, but you'll probably find at least one 220-volt circuit of use if you decide to go with a heavy bench grinder. You will need a 60-ampere, 220-volt circuit for many larger arc welders.

ACCESSORIES

Every welding shop will eventually need accessories. You will need tools like clamps and chipper scrapers. The *machinist's vise* is an important tool. A machinist's vise is expensive. You may get by with a mechanic's or bench vise. A bench vise will probably cost you about one-half or one-third of what a machinist's vise will with the same jaw opening. Both types are styled the same, with jaws and an anvil horn on one end. The bench vise isn't much good for extensive metal shaping on the anvil and anvil horn, as the metal will fatigue early. For extensive metal shaping, buy at least a small anvil, but for the occasional job the machinist's vise is a better buy.

The anvil is a handy tool for metal shaping. Anvils come in many weights and sizes. They can be found with plugs cut out to hold special shaping tools. You can use an old section of railroad track about 1½ feet long for an anvil. Lay the top of the rail on the surface to be used as a base, mark off four holes, and remove the metal. Cut the holes with your torch or drill them. Mount, and you have an anvil. It may not have the shape of the traditional anvil, but it will serve you well for a long time.

CODES

A careful check of any local code requirements for welding shops in residences should be made. You must be careful to attend to all local electrical and fire codes when setting up. If an inspection is made of the workshop, your certificate of occupancy for your home, if the shop is in the basement or an attached garage, could be rescinded if the codes are not met.

Follow through with whatever the codes tell you to do. Check with your local building inspector if you are confused or uncertain about any requirements.

WORKSHOP NEEDS

☐ **Ventilation:** probably the most important aspect from the standpoint of your personal health. Make certain the workshop area is or can be ventilated well enough to clear all fumes from the area.

☐ **Siting:** another major consideration.

☐ **Fireproofing:** if your workshop site is not fireproof, you will need to make it so or as nearly so as possible. In basement and garage workshops, it would be a good idea to fasten a sheet of fireproof material over the cutting and welding area.

☐ **Fire equipment:** you should always have a charged fire extinguisher of the dry chemical type. Use at least a five-pound model.

☐ **Separate work areas:** keep the different processes in the places where they belong. In other words, don't set up to cut metal a foot away from your bench grinder. Make sure that any flammable or easily destroyed material is separated from areas where torches or arc welders are in use.

☐ **Lights:** adequate fluorescent lighting of most work areas, with sufficient illumination over individual tools or areas.

☐ **Storage:** you will need enough storage space to keep your tools and materials out from underfoot and in some order which allows them to be easily located.

☐ **Access:** when siting your shop, consider supplies and the removal of finished work. Stairs and narrow doors present a problem when moving heavy metal objects.

☐ **Cleaning:** A good vacuum cleaner may not be essential, but it is a help. Make absolutely sure that any oil and grease spills are wiped up immediately and that no scrap or other material is underfoot.

Chapter 13

Propane and MAPP Gas Torches

These torches are not expensive and are handy for many applications. Generally, the most expensive propane torch will not cost more than $30. Propane and MAPP gas torches can not be used for fusion welding even thin metals. The better propane gas torches will provide good braze welding in light metals, and the MAPP gas torches will do slightly heavier work.

Propane and MAPP gas torches are single gas torches. In other words, no oxygen is added to the fuel gas as it comes from its storage cylinder. The fuel gas mixes only with air. The temperatures produced are thus lower than those with fuel gas torches, which accounts for the lessened capability.

There is only a single knob to open the fuel gas valve. Most torches light best if you turn the knob about one-eighth to one-quarter turn before snapping your spark lighter. Like larger oxyacetylene and oxy-MAPP torches, these single gas torches should not be lighted with a match (Figs. 13-1 through 13-3).

TORCH SELECTION

These small torches will help you solder pipe joints, braze weld a variety of parts, scrape paint, thaw pipes, and do a number of other jobs. While MAPP gas produces more heat and allows you to carry out slightly larger jobs, the torches

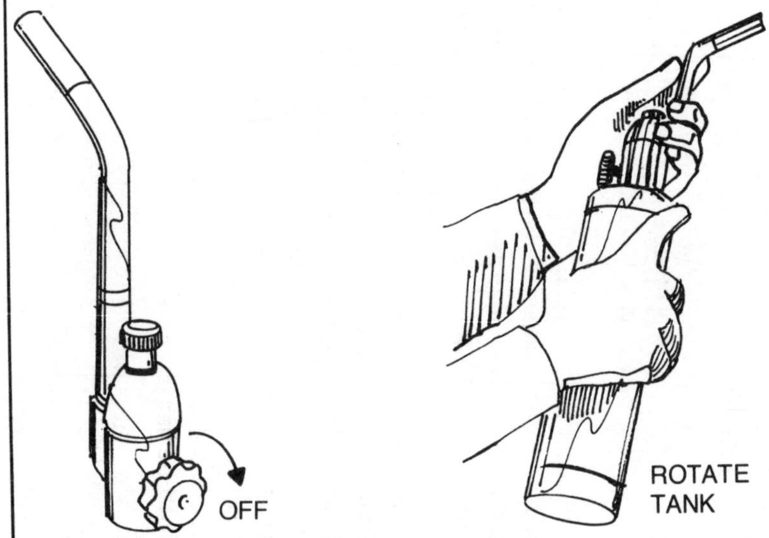

Fig. 13-1. Assembly instructions for single fuel gas torches.

ASSEMBLY INSTRUCTIONS

1. Be sure fuel valve is turned completely off. Turn fuel valve clockwise until snug.

2. Screw torch onto cylinder clockwise, by rotating tank.

ROTATE TANK

OFF

are usually about half as expensive as those designed to use propane. MAPP gas, or its equivalent by manufacturers other than Airco, is a more expensive fuel, costing from two to three times as much as propane. For some jobs like sweat soldering, MAPP gas produces almost too much heat.

You should never use MAPP gas in a torch designed solely for use with propane. The tips and orifices are not designed to handle the extra heat.

I would recommend that you buy a torch designed for use with MAPP gas if you expect even a moderate amount of medium sized work. In those cases where you don't need to use MAPP gas, the tank can be unscrewed and propane used instead to save money.

When selecting a torch, consider the range of tip styles to be found in that particular model. Some come with solid tips that hold the flame in back of the copper tip, which is then heated and used just like a soldering iron. Many torches have two or even three sizes of pencil flame tips, and many come with *flame spreaders*. A flame spreader is required when you are using the torch to soften paint for removal with a scraper.

PAINT REMOVAL

There are several safer ways of removing old paint than by using a propane torch. You can use paint scrapers and electric heaters. A propane torch should never be used to remove paint from enclosed walls of a house, barn, or other building. It is difficult or impossible to tell what exactly is behind such walls. If the house has been insulated with shredded paper or some other flammable substances, a smoldering fire could develop. It's best to use the propane torch on smaller items to be repainted or refinished. When used with care, you will get little or no scorching. To work, hold the flame on the surface material to be used, with the scraper held directly in back of the flame, and wait until the paint, lacquer, or other material bubbles. Use care in applying the heat. Though there may be little fire danger compared to wood surfaces, there is always a chance of heat distortion of lightweight metals.

LIGHTING INSTRUCTIONS

1. Turn the fuel valve ½ turn counterclockwise.
2. Ignite torch with sparker or hold match at lower edge of burner tube.

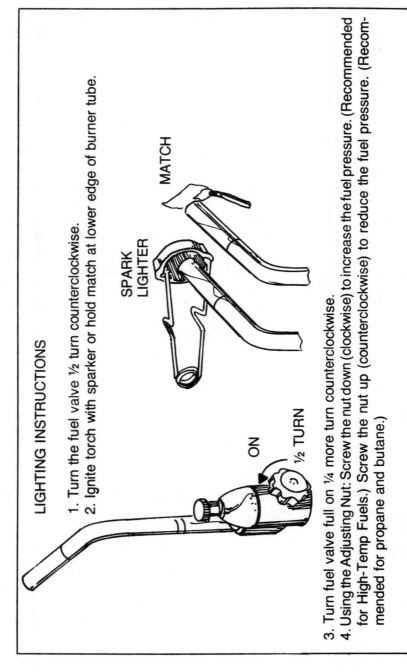

3. Turn fuel valve full on ¼ more turn counterclockwise.
4. Using the Adjusting Nut: Screw the nut down (clockwise) to increase the fuel pressure. (Recommended for High-Temp Fuels.) Screw the nut up (counterclockwise) to reduce the fuel pressure. (Recommended for propane and butane.)

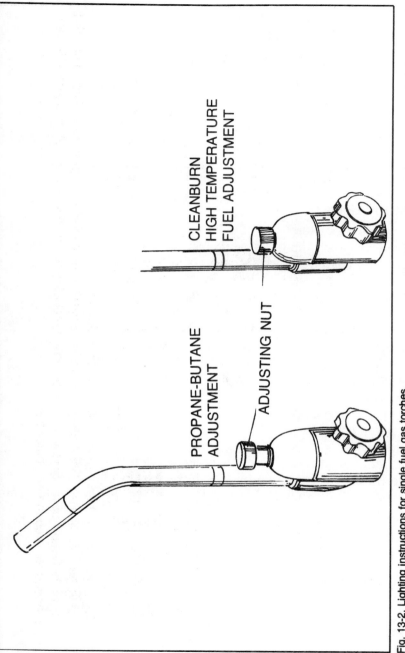

PROPANE-BUTANE
ADJUSTMENT

ADJUSTING NUT

CLEANBURN
HIGH TEMPERATURE
FUEL ADJUSTMENT

Fig. 13-2. Lighting instructions for single fuel gas torches.

261

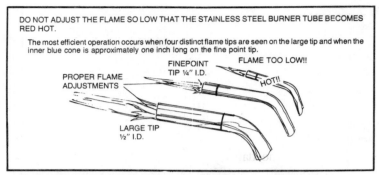

Fig. 13-3. Flame adjustment instructions for single fuel gas torches.

NUT REMOVAL

A propane torch or a MAPP torch can be invaluable when removing such things as old exhaust systems. Working in cramped quarters under my car, I couldn't get anywhere near enough leverage on the wrench handle to break loose the U-bolt nuts at the back end of the muffler and at the rear of the tailpipe. I applied heat for about 60 seconds to the tailpipe nuts in order to remove them. Because my stands weren't tall enough, I still had problems with the nuts at the back of the muffler, so I continued heating until the metal turned red. It was then possible to twist the nuts and the ends of the threaded U-bolt shafts right off.

Safety must always be considered when working around a car or pickup truck with a torch. First, if you have no lift, use jack stands or drive-on ramps. Second, make sure flammable material is not going to get overheated. Not too long ago I decided to increase my van's gas mileage by installing a dual exhaust. The van was old enough, and of the distinctly odd General Motors mid-engine design, to require a custom-built setup. Rather than try to haul the torch under the van, I borrowed a lift at a local service station. Unfortunately, the engine insulation was coated with grease to a degree I hadn't thought possible and caught fire. Only a fire extinguisher prevented a major problem. The insulation was fiberglass. I figured it might melt a bit, but I simply failed to consider the grease.

OTHER USES

There are many uses for propane torches. Last year a branch damaged a couple of shingles on my mother's house in late fall, after it had turned cool enough to make the material easy to crack if bent up or down (a necessary task when removing the damaged shingle and inserting the replacement). A light application of heat allowed me to lift an undamaged shingle high enough to pull the nails of the one underneath and slip the damaged one out. Use a wood block or heavy gloves to lift the good shingle. The new shingle is then slipped into place and nailed. The top shingle is pressed back into place with a dab of roofing cement under the edge.

Something had worn away some of the plating on the chimney flashing in a number of spots. Most modern flashing is aluminum and won't take standard leading procedures too well, so I would advise the use of roofing cement for repairs. This flashing was older, though far from old enough to be copper. A bar of solder, a torch, and some acid flux are needed. Simply apply the heat to the flashing. Spread the flux and then the bar solder over the worn spots.

Cleanup of flashing from repairs on homes is made easier using a putty scraper and heat from a propane torch. Roofing cement dries hard in a few days if the weather is cool, so it is difficult to remove from vent pipes and chimney surfaces. Heating the cement gently first makes scraping it off much easier.

When working on residences or barns, take care not to scorch any underlying materials that are flammable. Use the softest solder possible when soldering in such areas. The lower the melting point of the solder, the less chance any backing material has of meeting the 400 degree Fahrenheit charring point of underlying wood. Always work with a hose, extinguisher, or a pail or two of water at hand.

You will find other jobs to which you can apply your propane torch. It is one of the cheapest heat sources for working with metal and other materials, but one of the most versatile when extreme joint strength isn't essential to your job.

Glossary

abrasion—The effect on metal moved through materials that have a grinding action.

acetylene—A fuel gas used with oxygen to provide the hottest welding and cutting flame.

alignment—The holding of metals to be welded in place so that movement during heating isn't possible.

alloys—Elements added to basic metals to add to, or subtract from, natural properties (chromium added to iron to prevent oxidation).

alternating current—An electrical current which periodically alternates its direction of flow.

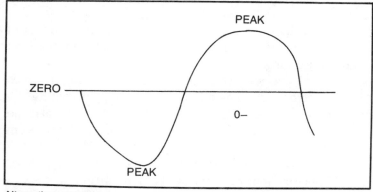

Alternating current.

aluminum—A lightweight, rapidly oxidizing metal with fast heat dissipating properties.

American Welding Society—The standard setting organization for welding supplies, techniques, and safety.

ammeter—A device for measuring the amount of current in a circuit.

annealing—A method of softening metal by applying specified degrees of heat.

arc—The hot band of sparks formed between the electrode and the base metal during arc welding.

arc blow—A magnetic disturbance of the arc causing it to move from an aimed path.

arc crater—The crater formed at the end of an arc-welded bead.

arc length—The distance from the end of the electrode to the point where the arc makes contact with the work surface.

arc welder—The machine used to provide power to make an arc. The person who is arc welding.

arc welding—The process of using an electrical arc to fuse two or more pieces of a metal.

argon—An inert gas used as an atmospheric contamination shield during tungsten inert gas welding.

backhand welding—A method of holding the torch and filler rod in oxyfuel gas welding in which the rod follows the torch tip.

backing strip—A metal strip, of an alloy similar to that of the base metal and filler metal, used to back up the joint during welding.

bare metal arc welding—A type of arc welding in which the electrode is uncoated or has a very light coating.

base metal—Often called parent metal, the base metal is the material being joined, whether with a weld or another form of jointure.

bevel—A type of edge preparation used to provide greater weld penetration, root fusion, and joint strength.

blowpipes—Another term for welding and cutting torches.

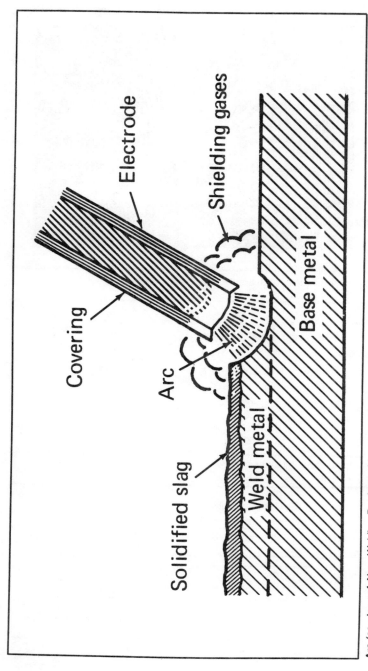

Arc (courtesy of Airco Welding Products, Murray Hill, NJ).

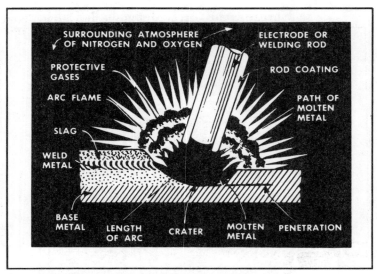

Arc crater (courtesy of Sears, Roebuck and Company).

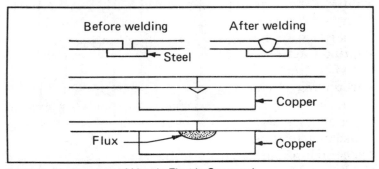

Backhand welding (courtesy of Airco Welding Products, Murray Hill, NJ).

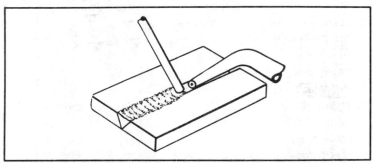

Backing strip (courtesy of Lincoln Electric Company).

braze weld—carbonizing (carburizing) flame

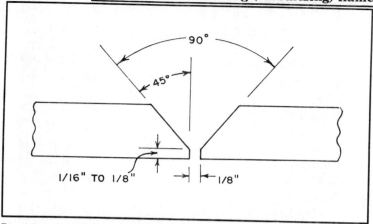

Bevel (courtesy of Airco Welding Products, Murray Hill, NJ).

braze weld—A form of nonfusion welding in which the base metal does not reach melting temperature.

brazing—A form of nonfusion jointure in which the base metal temperature is raised to at least 800 degrees Fahrenheit (427 centigrade).

brazing rod—A nonferrous filler rod used with braze welding and brazing.

butt joint—A frequently used welding and braze welding joint where the end or side planes of the base metal are usually joined with filler metal. In very light metals, filler metal may not be used.

cables—The parts of the arc welder which are attached to the work surface (the ground cable) and the welder, and to the electrode holder and the welder.

capillary action—Molecular attraction that causes a liquid to flow freely.

carbon arc—A type of arc that was originally often used for welding, using a nonconsumable carbon electrode and a rod for filler metal, that is now mostly used for braze welding and metal cutting.

carbonizing (carburizing) flame—A flame with a slight excess of fuel gas. It is often used when braze welding.

269

carbon steel—An alloy of iron with no more than 1.7 percent carbon. Plain carbon steel contains no other alloying elements.

cast iron—Iron, alloyed with at least 1.7 percent carbon, plus differing amounts of sulfur, phosphorous, manganese, and silicon.

chipping hammer—A slag removing tool.

chip testing—The process of using a cold chisel to chip off a piece of metal in order to see what type of metal it is.

clamps—Devices to help hold pieces in correct alignment.

conductivity, heat—The ability to dissipate heat.

copper—A metal resistant to corrosion and oxidation with excellent heat and electrical conductivity. Reddish brown in color, it has many alloys.

corner joint—A welding joint design in which two pieces of metal meet at an approximate 90-degree angle with no overlap.

corrosion resistance—The ability to resist the action of chemicals.

current—Electrical flow which is measured in amperes.

cutting—The process of using an oxyacetylene flame or an arc to cut metals.

cylinders—Containers for the compressed gases used in welding.

deposited metal—Metal deposited in the weld by a filler rod or electrode.

depth of fusion—The depth to which a weld fuses with the base metal, measured from the surface of the base metal to the point at which fusion stops.

desoldering—The process of heating and removing old solder so that new solder can be used on the connection.

direct current—An electrical current which always flows in one direction.

distortion—The twisting and buckling of metal caused by heat.

ductility—The ability of a metal to be bent or otherwise deformed without breaking.

SHIELDED METAL-ARC (MANUAL)

Position: Flat
Weld Quality Level: Commercial
Steel Weldability: Good
Welded From: One side

Weld Size, L (in.)	3/16	1/4	5/16	3/8	1/2
Plate Thickness (in.)	3/16	1/4	5/16	3/8	1/2
Pass	1	1	1	1	1 & 2
Electrode Class	E7024	E7024	E7024	E7024	E7024
Size	3/16	7/32	7/32	1/4	1/4
Current (amp) AC	250	320	350	400	410
Arc Speed (in./min)	21.0 — 25.0	18.0 — 22.0	14.5 — 17.5	13.0 — 16.0	11.5 — 14.5
Electrode Req'd (lb/ft)	0.101	0.133	0.198	0.240	0.530
Total Time (hr/ft of weld)	0.00870	0.0100	0.0125	0.0139	0.0308

Corner joint (courtesy of Lincoln Electric Company).

SHIELDED METAL-ARC (MANUAL)

Position: Flat
Weld Quality Level: Commercial
Steel Weldability: Good

18 – 10 ga

Plate Thickness (in.)	0.048 (18 ga)	0.060 (16 ga)	0.075 (14 ga)	0.105 (12 ga)	0.135 (10 ga)
Pass	1	1	1	1	1
Electrode Class	E6010	E6010	E6010	E6010	E6010
Size	3/32	1/8	1/8	5/32	3/16
Current (amp) DC(–)	50	80	85	115	140
Arc Speed (in/min)	45 – 50	43 – 48	40 – 45	40 – 45	37 – 42
Electrode Req'd (lb/ft)	0.0145	0.0232	0.0263	0.0382	0.0476
Total Time (hr/ft of weld)	0.00421	0.00439	0.00471	0.00471	0.00505

SHIELDED METAL-ARC (MANUAL)

Position: Vertical down
Weld Quality Level: Commercial
Steel Weldability: Good

Plate Thickness (in.)	0.048 (18 ga)	0.060 (16 ga)	0.075 (14 ga)	0.105 (12 ga)	0.135 (10 ga)
Pass	1	1	1	1	1
Electrode Class	E6010	E6010	E6010	E6010	E6010
Size	3/32	1/8	1/8	5/32	3/16
Current (amp) DC(−)	55	90	95	125	155
Arc Speed (in./min)	53 − 58	50 − 55	47 − 52	47 − 52	43 − 48
Electrode Req'd (lb/ft)	0.0141	0.0225	0.0251	0.0358	0.0473
Total Time (hr/ft of weld)	0.00361	0.00381	0.00404	0.00404	0.00439

Edge joints (courtesy of Lincoln Electric Company).

edge joint—Joints formed between the edges of two or more pieces of metal when those edges are parallel or close to it.

elasticity—The ability of a metal to return to its original shape after being deformed.

electrical conductivity—The ability of a metal to provide an electrical path and allow current flow.

electrical resistance—The resistance a metal offers to the flow of an electrical current.

electricity—Molecular motion that provides power.

electrode—A metal wire used as a terminal to form a complete circuit during arc welding, brazing, or cutting. Electrodes are classified as consumable (those that get used up rapidly and provide filler metal for the weld) and nonconsumable (those that don't get used up, except at a very slow rate, and provide nothing more than a terminal so that an arc can be formed).

electrode angle—The angle at which the electrode is held, which can be measured either from a vertical line or from the work surface.

electrode holder—An insulated clamplike device to hold electrodes during arc welding, cutting, and brazing.

electrode travel—The movement of the electrode along a joint.

fatigue strength—The ability of a metal to stand up to repeated stresses without breaking down.

ferrous metals—Metals containing iron.

fillet weld—A weld joining two surfaces at or near right angles to each other. A fillet weld across section is triangular or close to it.

flame hardening—A process in which an oxyacetylene flame is used to quickly heat a ferrous metal part, with quenching used to just as rapidly cool the part. Flame hardening allows the local hardening of parts in only those spots where hardness is necessary.

flange—A projecting rim on a piece of metal.

flat welding—Sometimes called downhand welding. Flat position welds are made from the upper side of the joint with the face of the weld approximately horizontal.

Fillet weld (courtesy of Airco Welding Products, Murray Hill, NJ).

Whip
first pass

Box weave
second pass

Straight weave

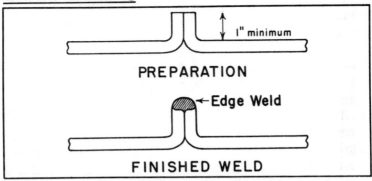

PREPARATION

←Edge Weld

FINISHED WELD

Flange (courtesy of Airco Welding Products, Murray Hill, NJ).

flux—Fusible chemicals used to dissolve and prevent the formation of oxides or other undesirable weld inclusions.

forehand welding—A method of holding the torch and welding rod in oxyfuel gas welding. The rod leads the torch tip.

fracture testing—Breaking a piece of metal in order to determine from its appearance what type of metal it is.

fuel gases—Gases used with oxygen to provide a flame for oxyfuel gas welding. The most common for the occasional welder are acetylene, MAPP gas, and propane.

fusion welding—Welding in which the base metals are fused—either to each other or to, or with, a filler metal.

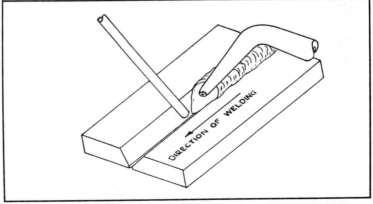

DIRECTION OF WELDING

Forehand welding (courtesy of Airco Welding Products, Murray Hill, NJ).

276

Metals to be fusion welded must be very similar in composition.

gas—The various gases used in welding process—both active, such as fuel gases and oxygen, and inert gases like argon.

gas flame—The flame produced by a fuel gas in combination with either air or oxygen.

goggles—An important safety feature during gas welding and brazing. Goggles provide protection against too intense light and possible hot metal spatters.

groove weld joints—Joints where a groove weld is made in the groove between two members being joined.

hardening—Several processes designed to increase the strength of steel or some other metals. In most, the metal is heated to a critical point and then cooled rapidly, though a few require slow cooling.

hardness—The ability of a metal to resist denting, abrasion, and penetration. There are several tests and methods of classifying metal hardness. Use either a steel ball or a diamond point to provide the indentation, which is then measured (with certain equipment, the indentation depth and the pressure needed are immediately converted on a dial on the test equipment). The *Brinell* test uses a steel ball and is limited to use on large surface areas. The Rockwell test can use either a steel ball or a diamond cone. There are nine different Rockwell scales, with the C scale the one most likely to be used by the occasional welder. The Vickers test uses a diamond point. The *Shore scleroscope* test gets its hardness classification by measuring the distance a diamond-tipped hammer rebounds from the metal surface.

heat source—The heat sources for welding of use to the occasional welder are oxygen-fuel gas mixtures and arc welders.

heat treating—Several methods of imparting greater strength to metals (see *flame hardening*).

SHIELDED METAL-ARC (MANUAL)
Special Procedures for ASTM A203 and A537 Steels

Position: Flat
Weld Quality Level: Code
Steel Weldability: Poor
Welded From: Two sides

	5/16		3/8	
Plate Thickness (in.)	5/16		3/8	
Pass	1 & 2	3 & 4*	1 – 3	4 – 6*
Electrode Class† Size	5/32	5/32	5/32	5/32
Current (amp) DC(+)	150	150	150	150
Arc Speed (in./min)	9 – 11	8 – 10	9 – 11	8 – 10
Electrode Req'd (lb/ft)	0.48		0.65	
Total Time (hr/ft of weld)	0.0844		0.127	
Interpass Temperature, Max. (°F)	150		150	

Diagram dimensions (5/16 plate): 5/32", 1/4", 5/16"
Diagram dimensions (3/8 plate): 3/16", 5/16", 3/8"

Position: Flat
Weld Quality Level: Code
Steel Weldability: Poor
Welded From: Two sides

Plate Thickness (in.)	1/2		5/8		3/4	
Pass	1 – 5	6 – 8*	1 – 7	8 – 10*	1 – 10	11 – 13*
Electrode Class†						
Size	5/32	5/32	5/32	5/32	5/32	5/32
Current (amp) DC(+)	150	150	150	150	150	150
Arc Speed (in./min)	7 – 9	8 – 10	7 – 9	8 – 10	7 – 9	8 – 10
Electrode Req'd (lb/ft)	1.40		1.79		2.25	
Total Time (hr/ft of weld)	0.188		0.238		0.313	
Interpass Temperature, Max. (°F)	175		200		225	

* Second side is gouged after first side is completed.
† See Tables 6-13 and 6-17.

Groove weld joints (courtesy of Lincoln Electric Company).

helium—A very light, inert gas sometimes used in tungsten inert gas welding.

horizontal position welding—A weld with two normal forms, one for groove welding and one for fillet welding. In groove welds, the weld will be made with its face in a vertical plane, with the weld running horizontally along that plane. In a fillet weld, the weld will be made on the upper side of a horizontal surface, while still being in contact with the vertical surface.

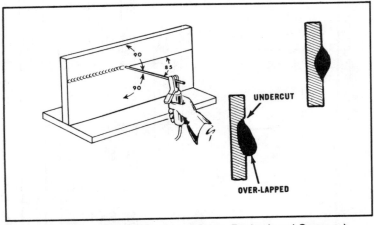

Horizontal position welding (courtesy of Sears, Roebuck and Company).

hoses—Hoses for welding are used for oxygen, fuel gases, and inert gases. Oxygen hoses are generally green, while fuel gas hoses are red. They must never be mixed up.

hose connectors—The connectors used for welding hoses will have right-hand threads for oxygen and left-hand threads for fuel gases, with a groove cut into the connectors for fuel gases to further help identification.

impact resistance—The ability of a metal to withstand impact without cracking or shattering (generally measured as toughness, with brittleness being the opposite). The ability of a metal to resist a sudden application of shock or force.

J-groove welds—Welds in which a J-shape has been ground in the groove to aid penetration and fusion.

joint—The point at which the welding takes place. There are five basic types of welding joints—the butt joint, the corner joint, the edge joint, the lap joint, and the tee joint.

keyholing—An aid to root fusion in some groove welds. The root of the weld is melted in a keyhole shape to help the flow of the filler metal.

lap joint—A joint formed by two overlapping members. It may be welded on one or both sides.

magnesium—A very light white metal with a low melting point. It resists corrosion well and oxidizes rapidly.

manual arc welding—The use of an arc welder and stick electrodes to manually weld several pieces of metal.

MAPP gas—Stabilized methyl acetylene-propadiene fuel gas often used in place of acetylene. MAPP is a trademark of Airco, Inc. Other companies market the same or very similar gas under various names. MAPP gas produces a temperature, when mixed with oxygen, of about 5300 degrees Fahrenheit (3089 degrees centigrade) and is safer to handle and work with than acetylene. It burns much hotter than propane does.

melting rate—The weight or the length of an electrode or welding rod that melts in a given unit of time.

metal electrode arc welding—See *manual arc welding*.

metal identification—Any of several ways of determining what alloy of a metal is on hand, so that correct welding rods or electrodes can be chosen. Primary methods of testing are visual, chip, fracture, and spark testing.

metal inert gas welding (MIG)—A shielded gas arc welding process in which a consumable wire electrode is used.

mixer—That part of a gas welding torch in which the oxygen and fuel gas are mixed.

SHIELDED METAL-ARC (MANUAL)

Position: Horizontal
Weld Quality Level: Commercial
Steel Weldability: Good

```
   |-- 18 - 10 ga
```

Plate Thickness (in.)	0.048 (18 ga)	0.060 (16 ga)	0.075 (14 ga)	0.105 (12 ga)	0.135 (10 ga)
Pass	1	1	1	1	1
Electrode Class	E6013	E6012	E6012	E6012	E6012
Size	3/32	1/8	5/32	3/16	3/16
Current (amp) DC(−)	70	105	145	200	210
Arc Speed (in./min)	19 − 23	21 − 26	20 − 24	18 − 22	14 − 18
Electrode Req'd (lb/ft)	0.0339	0.0427	0.0717	0.101	0.134
Total Time (hr/ft of weld)	0.00953	0.00851	0.00910	0.0100	0.0125

Lap joint (courtesy of Lincoln Electric Company).

Metal inert gas welding (MIG) (courtesy of Hobart Brothers Company).

overhead position welding—Welding that is done from the underside of the joint. It requires much practice and a special electrode (for arc welding).

overlap—A protrusion of metal beyond the bond at the toe of the weld.

oxygen—A colorless, tasteless gas essential for combustion and, for welding purposes, distributed at 99.5 percent purity. Oxygen is nonflammable, even though it supports combustion of other elements.

oxygen / fuel gas cutting—The process of using an oxyfuel gas flame to preheat metals to their kindling temperatures, at which point a stream of high velocity pure oxygen is introduced to sever the metal being cut.

parent metal—The same as base metal.

pass—A single weld bead deposit along a joint. The weld bead is the result. Several passes must be made in thick metals.

peening—Working of metal by rapid light hammer blows.

penetration—The distance that fusion takes place below the surface of metals being welded.

porosity—Gas pockets or other voids in metals.

postheating—Heat applied to the work after cutting, brazing, or welding.

pounds per square inch—Measurement used for gas pressures and for tensile strength—abbreviated psi.

preheating—Heat applied to the work before cutting, brazing, or welding to minimize thermal shock and slow the rate of cooling.

propane—A fuel gas with a limited heat output, mostly suitable for soft soldering and silver brazing. It has a tendency to give an oxidizing flame when used with oxygen. It burns in oxygen at 4600 degrees Fahrenheit (2538 degrees centigrade).

puddle—That portion of the weld metal that is molten at the point where heat is being applied.

quench—Rapid cooling of a metal by immersion in or dousing with water, brine, oil, or a blast of compressed air.

regulators—Gas controls and gauges used for metering oxygen, fuel gases, and inert gases.

reverse polarity—An arrangement of the arc welding leads so that the electrode is the positive pole and the workpiece is the negative pole. The electrode gets about 65 percent of the heat. Penetration is not as deep as with *straight polarity*. Abbreviated DCRP.

Rockwell test—A method of measuring metal hardness.

rods—Welding rods used as filler metal in tungsten inert gas welding and oxyfuel gas welding.

root of weld—The points at which the bottom of the weld intersect the base metal.

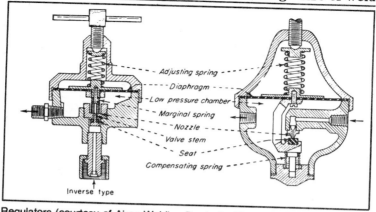

Regulators (courtesy of Airco Welding Products, Murray Hill, NJ).

shielded metal arc welding—Manual arc welding, also known as stick welding, in which the gases formed during welding protect the hot, molten puddle of weld metal from oxidation.

silver soldering—A soldering process in which the base metal stays below melting temperature. More properly known as silver brazing, the process is carried out at temperatures over 800 degrees Fahrenheit (427 degrees centigrade).

size of weld—For a groove weld, the size of weld is the depth of penetration. For equal fillet welds it is the leg length of the largest isosceles right triangle that can be formed within the weld cross section. For unequal fillet welds, the leg length of the largest right triangle that can be made in the weld cross section.

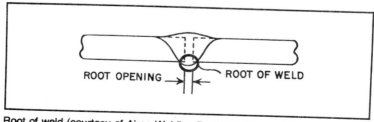

Root of weld (courtesy of Airco Welding Products, Murray Hill, NJ).

GROOVE WELD

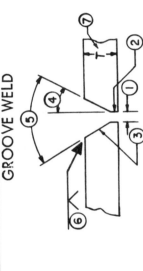

1. ROOT OPENING (RO): The separation between the members to be joined at the root of the joint.

2. ROOT FACE (RF): Groove face adjacent to the root of the joint.

3. GROOVE FACE: The surface of a member included in the groove.

4. BEVEL ANGLE (A): The angle formed between the prepared edge of a member and a plane perpendicular to the surface of the member.

5. GROOVE ANGLE (A): The total included angle of the groove between parts to be joined by a groove weld.

6. SIZE OF WELD(S): The joint penetration (depth of chamfering plus root penetration when specified).

7. PLATE THICKNESS (T) – Thickness of plate welded.

286

FILLET WELD

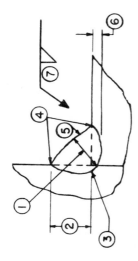

1. **THROAT OF A FILLET WELD:** The shortest distance from the root of the fillet weld to its face.

2. **LEG OF A FILLET WELD:** The distance from the root of the joint to the toe of the fillet weld.

3. **ROOT OF WELD:** Deepest point of useful penetration in a fillet weld.
4. **TOE OF A WELD:** The junction between the face of a weld and the base metal.
5. **FACE OF WELD:** The exposed surface of a weld on the side from which the welding was done.

6. **DEPTH OF FUSION:** The distance that fusion extends into the base metal.
7. **SIZE OF WELD(S):** Leg length of the fillet.

Size of weld (courtesy of Hobart Brothers Company).

slag—Nonmetallic material that forms on top of the weld.

slag inclusion—Nonmetallic material entrapped in the weld metal, thus weakening the joint.

soldering gun—An electric soldering tool shaped much like a pistol.

soldering iron—A soldering tool, either electric or heated by an outside source, with an insulated handle and a tip. It is used for melting solder and making connections.

soldering pencil—A miniature version of a *soldering iron.* It is used for soldering electronic components.

soldering torch—Any of a variety of open flame, handheld torches for heavier soldering. Fuel may be propane, butane, MAPP gas, or gasoline.

Soldering torch.

solders—Tin-lead alloys used to make joints where great strength is not needed. Melting point is under 800 degrees Fahrenheit (427 degrees centigrade).

spark testing—A method of determining what type of metal is being worked on. A sample of the metal is held against a powered grinding wheel, and the spark stream is checked for shape, length, color, and consistency of sparking.

spatter—Metal particles expelled during welding that do not form a part of the weld bead.

straight polarity—Arrangement of arc welding leads so that the workpiece is positive and the electrode is negative. The workpiece receives about 65 percent of the heat generated, and penetration is deep and narrow. Abbreviated DCSP.

stress relieving—Using heat and mechanical means to relieve internal stresses in metals after welding. Peening and heat tempering are two types.

stringer bead—A type of weld bead made with very little or no side to side movement of the rod or electrode.

striking an arc—The process by which electrical contact is made between the end of the electrode and the workpiece in order to start the arc for welding, cutting, or brazing.

sweat soldering—Used when solder must completely seal a joint to prevent air or water leakage. The heated joint draws, by capillary action, the molten solder in to seal the joint.

tack weld—A small weld to hold parts in alignment, often for further welding. With oxyfuel gas and gas shielded metal arc welding, tack welds are made without filler metal. With manual arc welding, some filler metal is deposited by the electrode.

tee joints—Joints formed between two pieces of metal placed at about right angles to each other in the shape of a "T."

temperature, liquidus—The lowest temperature at which an alloy becomes completely liquid.

Tee joints (courtesy of Lincoln Electric Company).

temperature, solidus—The highest temperature at which an alloy remains completely solid. An alloy with solidus and liquidus temperatures at the exact same point has no plastic range, but it has a definite melting point.

tempering—A heat treating process designed to improve physical properties, notably softness and toughness, of steels. Steel is hardened, cooled, and then reheated to a point slightly below the metal's critical point. It is held at that temperature for a specific length of time before being allowed to drop back to room temperature. The process also relieves internal metal stresses set up by the heat.

tensile strength—The degree to which metal will resist being torn apart—expressed in pounds per square inch (psi).

throat of weld—For a fillet weld, the shortest distance from the root of the weld to its face.

tinning—The process of coating the tip of a soldering iron, pencil, or gun with a thin film of solder. The precoating of metals prior to joining.

tips—Soldering tips for guns, irons, and pencils, as well as soldering torch tips, are interchangeable to accommodate a wide range of applications. Welding, cutting, and preheating torch tips are also available in many sizes.

toe of weld—The junction between the faces of the weld and the base metal.

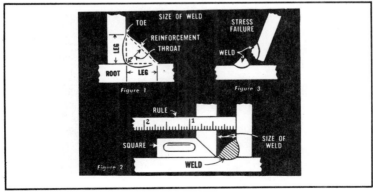

Toe of weld (courtesy of Sears, Roebuck and Company).

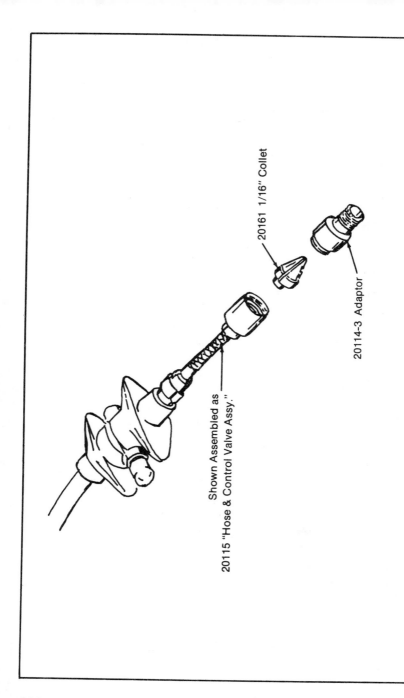

20161 1/16" Collet

20114-3 Adaptor

Shown Assembled as
20115 "Hose & Control Valve Assy."

292

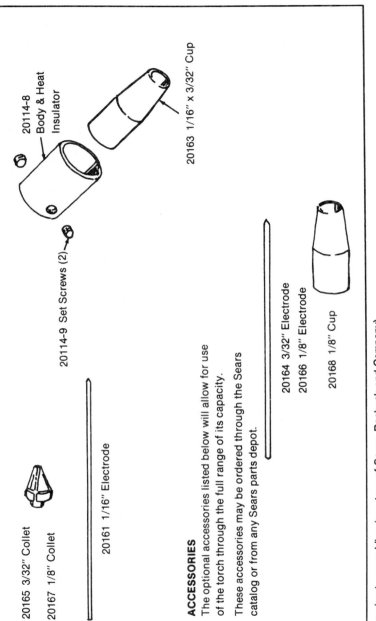

20165 3/32" Collet

20167 1/8" Collet

20161 1/16" Electrode

20114-9 Set Screws (2)

20114-8
Body & Heat
Insulator

20163 1/16" x 3/32" Cup

20164 3/32" Electrode

20166 1/8" Electrode

20168 1/8" Cup

ACCESSORIES

The optional accessories listed below will allow for use of the torch through the full range of its capacity.

These accessories may be ordered through the Sears catalog or from any Sears parts depot.

Tungsten inert gas welding (courtesy of Sears, Roebuck and Company).

torch—A welding torch consists of several parts, from the back to the body and the valves, as well as an internal mixing chamber or mixer and gas-carrying tubes. A cutting torch is similar but has a lever, plus an extra gas tube, to allow the operator to use a stream of pure oxygen when needed.

toughness—The ability of a metal to resist sudden shock.

tungsten electrode—A tungsten wire electrode, classified as nonconsumable, used to provide an arc for *tungsten inert gas welding.* It does not provide filler metal for the weld.

tungsten inert gas welding—An arc welding process that uses a tungsten electrode to form an arc which is then shielded by an inert gas. The inert gas is usually argon for home and farm applications. This type of welding is exceptionally useful on metals such as aluminum which are hard to weld because of rapid oxide formation.

U-groove—A groove ground in a butt joint, in the shape of a U, to improve penetration and joint strength and reduce the amount of filler metal required.

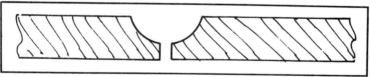

U-groove.

undercut—An undesirable groove that forms on the base metal adjacent to the toe of the weld and remains unfilled with filler metal.

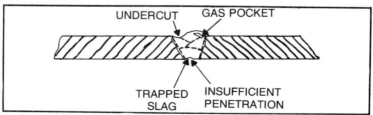

Undercut (courtesy of Sears, Roebuck and Company).

vertical welding—Welding in a position where the weld bead runs vertically.

V-groove—A groove ground in a butt joint to aid penetration and fusion.

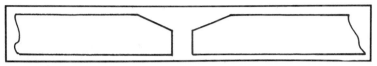

V-groove.

weaving—A welding technique while the rod or electrode is being oscillated.

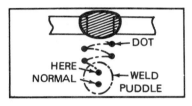

Weaving (courtesy of Hobart Brothers Company).

weld—A localized coalescence of metals where the filler metal has approximately the same melting temperature as the base metal, or where the filler metal has a lower melting temperature than the base metal(s), but greater than 800 degrees Fahrenheit (427 degrees centigrade).

weldability—The ability of a metal to respond to a welding method.

welding—The metal joining process used to form welded joints.

welding helmet—A face and eye protecting device with dark, interchangeable lenses.

weldment—The assembly of parts joined by welding.

weld metal—The portion of a weld that has been melted during the welding process.

wetting—The spreading of a liquid filler metal or flux on the parent metal.

wiped joint—A joint made with solder having a wide plastic range. The solder is manipulated with a handheld cloth as required.

Appendix A
Arc and Gas
Welding Safety Procedures

☐ Make sure all arc welding equipment is installed according to manufacturer's instructions and local and national codes. It must be in good working condition.

☐ Wear clothing suitable to the work being done. Avoid exposing your hands, face, and eyes to torch heat and arc radiation. Avoid wearing easily melted synthetics.

☐ Wear correct eye protection for the process being used.

☐ Keep volatile, flammable materials out of the work area, or protect flammables with asbestos shielding.

☐ Keep the work area clean and free of underfoot hazards.

☐ Use care when handling compressed gas cylinders.

☐ Secure compressed gas cylinders to a solid support with chains or use a suitable hand truck and chain.

☐ Do not weld in a confined space without taking special precautions for ventilation and volatile or flammable materials.

☐ Take special precautions when welding on containers that *might* have contained volatile materials.

☐ Do not weld on sealed containers or pressurized containers of any kind.

☐ Use mechanical exhaust means and wear a respirator when welding or cutting lead, cadmium, chrominum, manganese, bronze, zinc, or galvanized materials.

☐ Wear special boots and use an insulated platform when arc welding in damp conditions.

☐ Do not use welding cables with frayed, abraded, or cracked insulation.

☐ Install and use a bracket for the electrode holder.

☐ Dispose of electrode and rod stubs in a metal container as even short stubs can be a hazard underfoot.

☐ Make sure any observers watching you weld have suitable eye and face protection.

☐ If you must work above the ground, make sure that the scaffold or other work surface is solid. Wear a lifeline where there are no safety railings.

☐ When flame cutting, sparks and molten metal can fly as far as 40 feet. Make sure hoses, regulators, and cylinders are not in line to be hit, and that other materials needing protection are out of the way or covered with flame proof material.

☐ Use the correct torches and tips for the fuel gases being used.

☐ Never use oxygen as compressed air after the original blowing out of talc from the hoses.

☐ Never use acetylene at pressures above 15 pounds per square inch.

☐ Never allow oil or grease to come in contact with gas welding regulators, connectors, or torches.

☐ Use a minimum of force when making connections in gas welding equipment. Use Ivory detergent and water to check for gas leaks.

☐ Crack cylinder valves before attaching regulators so that any foreign material will be cleared from the seating surfaces.

☐ Use the correct lighting sequence for oxyfuel gas torches. If there is any doubt, always follow the manufacturer's instructions.

    —Open the acetylene valve slightly.

    —Open the torch acetylene valve a quarter turn.

    —Screw in the acetylene regulator valve adjusting handle to get working pressure.

    —Shut down the acetylene torch valve. The purge is complete.

    —Repeat the previous four steps with the oxygen cylinder regulator and torch.

—Open the acetylene torch valve about a quarter turn and light the torch with a spark lighter (never a match).

—Open the oxygen torch valve a partial turn.

—Adjust to a neutral flame.

☐ Always use the correct procedure and sequence when shutting off an oxyfuel gas outfit.

—Close the acetylene torch valve.

—Close the oxygen torch valve.

—Close the cylinder valves—acetylene first.

—Open the acetylene torch valve.

—Open the oxygen torch valve.

—Close both torch valves.

—Back off on the regulator adjusting valve handles until you can feel no pressure.

Appendix B
Welding Problems

distortion

WHY
1. Improper tack welding and/or faulty joint preparation
2. Improper bead sequence
3. Improper set-up and fixturing
4. Excessive weld size

WHAT TO DO
1. Tack weld parts with allowance for distortion

2. Use proper bead sequencing
3. Tack or clamp parts securely
4. Make welds to specified size

spatter

WHY
1. Arc blow
2. Welding current too high
3. Too long an arc length
4. Wet, unclean or damaged electrode

WHAT TO DO
1. Attempt to reduce the effect of arc blow
2. Reduce welding current

3. Reduce arc length
4. Properly maintain and store electrodes

lack of fusion

WHY
1. Improper travel speed
2. Welding current too low
3. Faulty joint preparation
4. Too large an electrode diameter
5. Magnetic arc blow
6. Wrong electrode angle

WHAT TO DO
1. Reduce travel speed
2. Increase welding current
3. Weld design should allow

electrode accessibility to all surfaces within the joint
4. Reduce electrode diameter
5. Reduce effects of magnetic arc blow
6. Use proper electrode angles

Fig. B-1. Causes and cures of common welding problems (courtesy of Hobart Brothers Company) (continued from page 301).

overlapping

WHY
1. Too slow travel speed
2. Incorrect electrode angle
3. Too large an electrode

WHAT TO DO
1. Increase travel speed
2. Use proper electrode angles
3. Use a smaller electrode size

poor penetration

WHY
1. Travel speed too fast
2. Welding current too low
3. Poor joint design and/or preparation
4. Electrode diameter too large
5. Wrong type of electrode
6. Excessively long arc length

WHAT TO DO
1. Decrease travel speed
2. Increase welding current
3. Increase root opening or decrease root face

4. Use smaller electrode
5. Use electrode w/deeper penetration characteristics
6. Reduce arc length

magnetic arc blow

WHY
1. Unbalanced magnetic field during welding
2. Excessive magnetism in parts or fixture

WHAT TO DO
1. Use alternating current
2. Reduce welding current and arc length

3. Change the location of the work connection on the workpiece

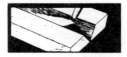

inclusion

WHY
1. Incomplete slag removal between passes
2. Erratic travel speed
3. Too wide a weaving motion
4. Too large an electrode
5. Letting slag run ahead of arc
6. Tungsten spitting or sticking

WHAT TO DO
1. Completely remove slag between passes
2. Use a uniform travel speed
3. Reduce width of weaving technique

4. Use a smaller electrode size for better access to joint
5. Increase travel speed or change electrode angle or reduce arc length
6. Properly prepare tungsten and use proper current

porous welds

WHY
1. Excessively long or short arc length
2. Welding current too high
3. Insufficient or damp shielding gas
4. Too fast travel speed
5. Base metal surface covered with oil, grease, moisture, rust, mill scale, etc.
6. Wet, unclean or damaged electrode

WHAT TO DO
1. Maintain proper arc length

2. Use proper welding current
3. Increase Gas Flowrate & check gas purity
4. Reduce travel speed
5. Properly clean base metal prior to welding
6. Properly maintain and store electrode

cracked welds

WHY
1. Insufficient weld size
2. Excessive joint restraint
3. Poor joint design and/or preparation
4. Filler metal does not match base metal
5. Rapid cooling rate
6. Base metal surface covered with oil, grease, moisture, rust, dirt or mill scale

WHAT TO DO
1. Adjust weld size to part thickness

2. Reduce joint restraint through proper design
3. Select the proper joint design
4. Use more ductile filler
5. Reduce cooling rate through preheat
6. Properly clean base metal prior to welding

undercutting

WHY
1. Faulty electrode manipulation
2. Welding current too high
3. Too long an arc length
4. Too fast travel speed
5. Arc blow

WHAT TO DO
1. Pause at each side of the weld bead when using a weaving technique
2. Use proper electrode angles

3. Use proper welding current for electrode size and welding position
4. Reduce arc length
5. Reduce travel speed
6. Reduce effects of arc blow

Fig. B-1. Causes and cures of common welding problems (courtesy of Hobart Brothers Company) (continued from page 301).

Appendix C
Welding Project

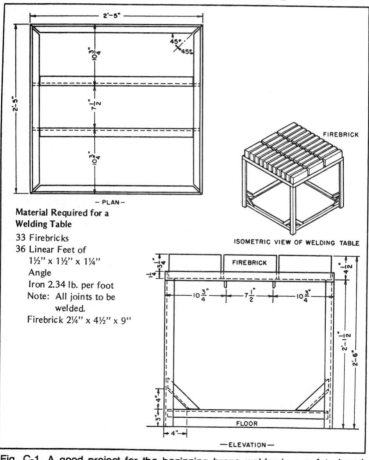

– PLAN –

Material Required for a Welding Table

33 Firebricks
36 Linear Feet of
1½" x 1½" x 1¼"
Angle
Iron 2.34 lb. per foot
Note: All joints to be
welded.
Firebrick 2¼" x 4½" x 9"

FIREBRICK

ISOMETRIC VIEW OF WELDING TABLE

FIREBRICK

FLOOR

– ELEVATION –

Fig. C-1. A good project for the beginning braze welder is a safety bench (courtesy of Airco Welding Products, Murray Hill, NJ).

Appendix D
Charts and Tables

TEMPERATURE CONVERSION CHART

TEMPERATURES:

Temperatures expressed
as DEGREES CELSIUS:

°F	°C
425	219
1050	566
1090	588
1125	607
1130	610
1400	760
1600	871
2000	1093

°C	°F
1000	1800
950	1700
900	1600
850	
800	1500
750	1400
700	1300
650	1200
600	1100
550	1000
500	900
450	800
400	700
350	
300	600
250	500
200	400
150	300
100	200
50	100
0	0
-50	

Fig. D-1. Temperature conversion information (courtesy of Sears, Roebuck and Company).

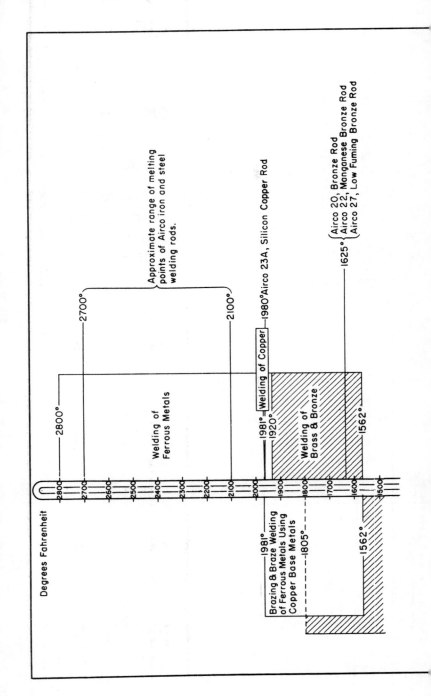

Degrees Fahrenheit

Approximate range of melting points of Airco iron and steel welding rods.

2800°

2700°

2100°

1980° Airco 23A, Silicon Copper Rod

1981° Welding of Copper

1920°

Welding of Ferrous Metals

Welding of Brass & Bronze

1625° { Airco 20, Bronze Rod
Airco 22, Manganese Bronze Rod
Airco 27, Low Fuming Bronze Rod

1562°

1981° Brazing & Braze Welding of Ferrous Metals Using Copper Base Metals

1805°

1562°

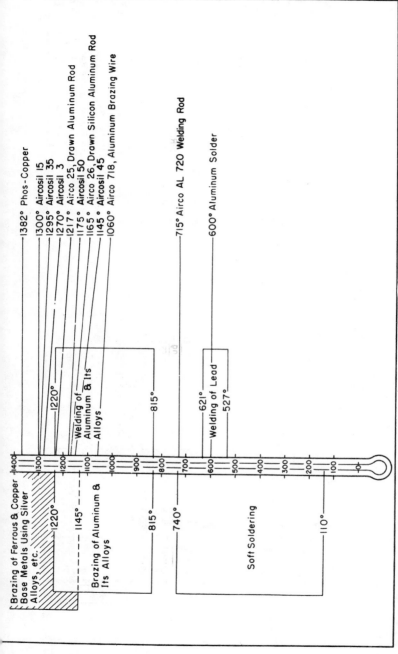

Fig. D-2. Chart of welding processes and filler rods as related to temperatures (courtesy of Airco Welding Products, Murray Hill, NJ).

Table D-1. MAPP Gas Properties (courtesy of Airco Welding Products, Murray Hill, NJ).

	MAPP gas	acetylene	natural gas	propane
SAFETY				
Shock sensitivity	Stable	Unstable	Stable	Stable
Explosive limits in oxygen, %	2.5-60	3.0-93	5.0-59	2.4-57
Explosive limits in air, %	3.4-10.8	2.5-80	5.3-14	2.3-9-5
Maximum allowable regulator pressure, psi	Cylinder	15	Line	Cylinder
Burning velocity in oxygen, ft /sec	15.4	22.7	15.2	12.2
Tendency to backfire	Slight	Considerable	Slight	Slight
Toxicity	Low	Low	Low	Low
Reactions with common materials	Avoid alloys with more than 67% copper	Avoid alloys with more than 67% copper	Few restrictions	Few restrictions
PHYSICAL PROPERTIES				
Specific gravity of liquid (60 /60°F.)	0.576	—	—	0.507
Pounds per gallon liquid at 60°F.	4.80	—	—	4.28
Cubic feet per pound of gas at 60°F.	8.85	14.6	23.6	8.66
Specific gravity of gas (air = 1) at 60° F.	1.48	0.906	0.62	1.52
Vapor pressure at 70°F., psig	94	—	—	120
Boiling range, °F. 760 mm. Hg	−36 to −4	−84	−161	−50
Flame temperature in oxygen, °F.	5,301	5,589	4,600	4,579
Latent heat of vaporization at 25°C., BTU /lb	227	—	—	184
Total heating value (after vaporization) BTU /lb	21,100	21,500	23,900	21,800

Table D-2. Heating Values of Fuel Gases (courtesy of Airco Welding Products, Murray Hill, NJ).

fuel	flame temp. (°F.)	primary flame (BTU /cu. ft.)	secondary flame (BTU /cu. ft.)	total heat (BTU /cu. ft.)
MAPP Gas	5301	571	1889	2406
Acetylene	5589	507	963	1470
Propane	4579	255	2243	2498
Natural Gas	4600	11	989	1000

Table D-3. Troubleshooting Chart (courtesy of Sears, Roebuck and Company).

trouble	probable cause	remedy
Welding Tip popping	Too close to work	Move further from work area
Flames not clearly defined, smooth or even	Dirty tip	Clean with tip cleaner or replace tip
Cutting Tip popping	Too loose Nicked seat	Tighten tip nut Replace tip
Leak around needle valve	Packing nut loose	Snug packing nut
Difficult to light	Needle valve open too wide	Partly close needle valve
Flame change when cutting	Oxygen or acetylene cylinder almost empty	Replace cylinder with full one

Table D-4. Melting Points of Metals.

metal or alloy	melting point, °F
Aluminum, Pure	1218
Brass and Bronze	1600-1660
Copper	1981
Iron, Cast and Malleable	2300
Lead, Pure	620
Magnesium	1240
Monel	2400
Nickel	2646
Silver, Pure	1762
Steel, Hi-Carbon (0.40% to 0.70% Carbon)	2500
Steel, Medium Carbon (0.15% to 0.40% Carbon)	2600
Steel, Low Carbon (less than 0.15%)	2700
Stainless Steel, 18% Chromium, 8% Nickel	2550
Titanium	3270
Tungsten	6152
Zinc, Cast or Rolled	786

Table D-5. Suggested Gas Welding Procedures for Nonferrous Metals (courtesy of Airco Welding Products, Murray Hill, NJ).

BASE METAL	GAS PROCESS*	FLAME TYPE
Aluminum Alloys 2S & 3S	Welding	Neutral
	Brazing	Neutral
Aluminum Alloys 52S, 53S, 61S, & 63S	Welding	Neutral
	Brazing	Neutral
Brass, Red	Welding	Oxidizing
Brass, Yellow	Welding	Oxidizing
Bronze, Aluminum (Below 5% Al.)	Welding	Slightly Carburizing
Bronze, Phosphor	Welding	Neutral
	Braze Welding	Oxidizing
Copper, Beryllium	Brazing	Slightly Carburizing
Copper, Deoxidized & Copper, Electrolytic	Welding	Slightly Oxidizing or Neutral
	Braze Welding	Slightly Oxidizing
Muntz Metal	Welding	Slightly Oxidizing
Nickel & High Nickel Alloys	Welding	Neutral or Slightly Carburizing
Nickel Silver	Braze Welding	Neutral

*All of the metals in the table with the exception of the Aluminum alloys can be brazed using either the silver or copper base brazing alloys. The melting point and composition of the base metal will determine the correct brazing alloy.

FILLER METAL	FLUX	REMARKS
Airco No. 25	Airco Napolitan	
Airco No. 26 or 718	Airco Elite	
Airco No. 26	Airco Napolitan	
Airco 718	Airco Elite	Recommend etch with acid before brazing
Airco No. 20, 22, or 27**	Airco Hi-Test or Marvel	
Airco No. 20	Airco Hi-Test or Marvel	
Match base metal	Special Flux	Weld as continuously as possible. Don't weld Aluminum Bronze with above 5% Al.
Match base metal (Grade E phosphor bronze)	Airco Hi-Test or Marvel	
Airco No. 20, 22, or 27	Airco Hi-Test or Marvel	
Aircosil 50, 45, 35, or 3	Aircosil Flux	Only process recommended
Airco No. 23A	Airco Marvel	
Airco No. 20, 22, or 27	Airco Marvel	
Airco No. 20, 22, or 27	Airco Hi-Test or Marvel	Airco No. 22 especially recommended for manganese bronze
Match base metal	Manufacturer's recommendation	
Airco No. 21	Airco Hi-Test or Marvel	

**Note: Certain corrosive conditions do not permit the use of high zinc brass filler metals; for those cases weld with silicon copper rods.*

BASE METAL	GAS PROCESS*	FLAME TYPE
Cast Iron, Gray	Welding	Neutral
	Braze Welding	Neutral
Cast Iron, Malleable	Braze Welding	Slightly Oxidizing
Cast Iron, Moly	Welding	Neutral
Galvanized Iron or Steel	Braze Welding	
Steel, Cast	Welding	Slightly Carburizing
	Braze Welding	Neutral
Steel, High Carbon (0.45% and up)	Welding	Carburizing
Steel, Low Carbon (up to 0.30%)	Welding	Slightly Carburizing
Steel, Medium Carbon (0.30-0.45%)	Welding	Slightly Carburizing
Steel, Miscellaneous Alloy	Welding	Slightly Carburizing
Steel, Stainless	Welding	Slightly Carburizing
Wrought Iron	Welding	Neutral
	Braze Welding	Neutral

*All of these metals can be brazed using the Easy-Flo's or Airco
Nos. 20, 22, or 27.

ROD	FLUX	REMARKS
Airco No. 9	Airco Atlas	Preheat (900° F. min.)
Airco No. 20	Airco Hi-Test, Hi-Bond, or Marvel	Slight preheat
Airco No. 20	Airco Hi-Test, Hi-Bond, or Marvel	
Airco No. 10	Airco Atlas	Preheat (900° F. min.) Postheat (1100° F. min.)
Airco No. 22 or No. 27	Airco Hi-Test	Provide proper ventilation to get rid of zinc fumes
Airco No. 4	None	
Airco No. 22 or No. 27	Airco Hi-Test, Hi-Bond, or Marvel	
Airco No. 1 or No. 4**	None	
Airco No. 1, 4, or 7	None	
Airco No. 1 or No. 4	None	
Special rods	None	
Rod matching base metal	Airco Stainless Steel Flux	
Airco No. 1, 4, or 7	None	
Airco No. 20, 22, or 27	Airco Hi-Test Hi-Bond, or Marvel	

**Note: Satisfactory for some conditions, but other conditions will require a high carbon filler metal.*

Table D-7. Variations in Oxygen Cylinder Pressures with Temperature Changes.

Gauge pressures indicated for varying temperature conditions on a full cylinder initially charged to 2200 psi at 70°F. Values identical for 244 cu. ft. and 122 cu. ft. cylinder.

temperature degrees F.	pressure psi approx.	temperature degrees F.	pressure psi approx.
120	2500	30	1960
100	2380	20	1900
80	2260	10	1840
70	2200	0	1780
60	2140	−10	1720
50	2080	−20	1660
40	2020		

Table D-8. Oxygen Cylinder Content.

Indicated by Gauge Pressure at 70° F. 244 Cu. Ft. Cylinder

gauge pressure psi	content cu. ft.	gauge pressure psi	content cu. ft.
190	20	1200	130
285	30	1285	140
380	40	1375	150
475	50	1465	160
565	60	1550	170
655	70	1640	180
745	80	1730	190
840	90	1820	200
930	100	1910	210
1020	110	2000	220
1110	120	2090	230
		2200	244

122-cu. ft. cylinder content one-half above volumes.

Appendix E
Good and Bad Beads

GOOD Proper Current Voltage & Speed	BAD Welding Current Too Low	BAD Welding Current Too High

Cross-section Weld Bead	*Cross-section Weld Bead*	*Cross-section Weld Bead*

 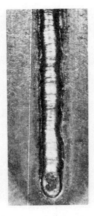 

Face Weld Bead	*Face Weld Bead*	*Face Weld Bead*
A smooth, regular, well formed bead.	Excessive piling up of weld metal.	Excessive spatter to be cleaned off.
No undercutting, over-lapping or piling up.	Overlapping bead has poor penetration.	Undercutting along edges weakens joint.
Uniform in cross section.	Slow up progress.	Irregular deposit.
Excellent weld at minimum material and labor cost.	Wasted electrodes and productive time.	Wasted electrodes and productive time.

Fig. E-1. Shielded metal arc welding (courtesy of Hobart Brothers Company).

BAD **Arc Too Long** **(Voltage Too High)**	**BAD** **Welding Speed** **Too Fast**	**BAD** **Welding Speed** **Too Slow**

Cross-section Weld Bead	*Cross-section Weld Bead*	*Cross-section Weld Bead*

Face Weld Bead	*Face Weld Bead*	*Face Weld Bead*
Bead very irregular with poor penetration.	Bead too small, with contour irregular.	Excessive piling up of weld metal.
Weld metal not properly shielded.	Not enough weld metal in the cross section.	Overlapping without penetration at edges.
An inefficient weld.	Weld not strong enough.	Too much time consumed.
Wasted electrodes and productive time	Wasted electrodes and productive time.	Wasted electrodes and productive time.

GOOD Proper Current Voltage & Travel	BAD Welding Current Too Low (High Voltage)	BAD Welding Current Too High (Low Voltage)
Cross-section Fillet	*Cross-section Fillet*	*Cross-section Fillet*
Cross-section Weld Bead	*Cross-section Weld Bead*	*Cross-section Weld Bead*
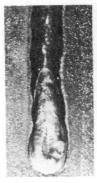		
Face Weld Bead	*Face Weld Bead*	*Face Weld Bead*
Smooth, regular, well formed bead.	Excessive spatter and porosity.	Weld bead excessively convex and narrow.
No undercut, overlap, or pileup.	Weld bead excessively wide and flat.	Difficult slag removal.
Uniform in cross section.	Undercutting along edges weakens joint.	Wasted filler metal and productive time.
Excellent weld at minimum material and labor cost.	Irregular bead contour.	

Fig. E-2. FabCO—with external shielding gas (courtesy of Hobart Brothers Company).

320

BAD **Welding Speed** **Too Fast**	**BAD** **Welding Speed** **Too Slow**	**BAD** **Insufficient Shielding** **Gas Coverage**

Cross-section Fillet

Cross-section Fillet

Cross-section Fillet

Cross-section Weld Bead

Cross-section Weld Bead

Cross-section Weld Bead

Face Weld Bead

Face Weld Bead

Face Weld Bead

Bead too small with irregular contour.	Excessive bead widths.	Weld very porous and brittle.
Not enough weld metal in cross section.	Overlapping without penetration at edges.	Little weld strength.
Poor mechanical properties.	Fillet with unequal legs.	
Undercut at toe lines of fillet.	Wasted filler metal and productive time.	

GOOD Proper Current Voltage & Travel	BAD Welding Current Too Low (High Voltage)	BAD Welding Current Too High (Low Voltage)
Cross-section Fillet	*Cross-section Fillet*	*Cross-section Fillet*
Cross-section Weld Bead	*Cross-section Weld Bead*	*Cross-section Weld Bead*
Face Weld Bead	*Face Weld Bead*	*Face Weld Bead*
Smooth, regular, well formed weld bead. No undercut, overlap, or pileup. Uniform in cross section. Excellent weld at minimum material and labor cost.	Excessive spatter and porosity. Weld bead excessively wide and flat. Undercutting along edges weakens joint. Irregular bead contour.	Weld bead excessively convex and narrow. Difficult slag removal. Wasted filler metal and productive time.

Fig. E-3. Fabshield—without external shielding gas (courtesy of Hobart Brothers Company).

BAD Welding Speed Too Fast	**BAD** Welding Speed Too Slow	**BAD** Insufficient Shielding Gas Coverage

Cross-section Fillet

Cross-section Fillet

Cross-section Fillet

Cross-section Weld Bead Cross-section Weld Bead Cross-section Weld Bead

Face Weld Bead	*Face Weld Bead*	*Face Weld Bead*
Bead too small with irregular contour.	Excessive bead width.	Excessive spatter and porosity.
Not enough weld metal in cross section.	Overlapping without penetration at edges.	Bead very irregular with poor penetration.
Poor mechanical properties.	Fillet with unequal legs.	Weld metal not properly shielded.
Undercut at toe lines of fillet.	Wasted filler metal and productive time.	Wasted electrode and productive time.

Index